Managing people

ENGINEERING MANAGEMENT

Series Editor S. H. Wearne, BSc(Eng), PhD, FICE, FBIM, Consultant, Director of Institution courses and in-company training

Editorial panel D. E. Neale, FICE; D. P. Maguire, BSc, FICE; D. J. Ricketts, BSc; B. A. O. Hewett, BSc(Eng), MSc, FICE; J. V. Tagg, FICE

Other titles in the series

Civil engineering insurance and bonding, P. Madge

Marketing of engineering services, J. B. H. Scanlon

Construction planning, R. H. Neale and D. E. Neale

Civil engineering contracts, S. H. Wearne

Management of design offices, P. A. Rutter (Ed)

Project evaluation, R. K. Corrie (Ed)

Financial control, N. M. L. Barnes (Ed)

Control of engineering projects, S. H. Wearne *et al.*

ENGINEERING MANAGEMENT

Managing people

Edited by
Stanley Martin, MSc, FICE, FIHT, FBIM
Fred Grover, BSc(Eng), FICE, ACGI, FIHT

Thomas Telford Ltd, London

Published by Thomas Telford Ltd, Thomas Telford House,
1 Heron Quay, London E14 9XF

First published 1988

British Library Cataloguing in Publication Data
Martin, A.S. and Grover, F. (Eds)
   Managing people — (Engineering management)
   1. Construction industries. Personnel management.
   I. Martin, A.S.   II. Grover, F.   III. Series
   624'.068'3

ISBN   0 7277 1354 X

Typeset in Great Britain by MHL Typesetting Limited, Coventry
Printed and bound in Great Britain by Billing and Sons Limited,
Worcester

Foreword

Construction is a labour and management intensive industry responsible for some 10% of the GNP and employing nearly $1\frac{1}{2}$ million people. Construction organizations are only as strong as their skills and management expertise, and it is vital that these are kept at the highest level so that the organization may flourish and grow in a competitive environment.

Those who are entrusted with a management function within the industry have to develop a wide range of skills which will enable them to maximize the benefit to their shareholders or to the public in the long-term. They need to optimize the application of resources, be they fixed assets of people's skills. The ability to manage and motivate people is of the utmost important, requiring communication skills, leadership qualities and the ability to create confidence and credibility.

This book deals with the broad range of subjects and skills that are required and provides sound advice for those who are seeking a position of managerial responsibility in the construction industry and wish to ensure their success. It is about the management of people and is aimed at civil engineers who are taking on management responsibilities for the first time.

H.W.A. Francis,   15 June 1988
President, Institution of Civil Engineers.

Acknowledgements

We are grateful to several of our colleagues for their contributions to the text and particularly to D.A. Barratt for acting as co-ordinator and compiler of Chapter 5, and to E. Hook and J.T. Pike for supplying two of the organization charts for Chapter 3.

Preface

The need for management training
This book is about the management of people, arguably the most important resource in any organization.

It is aimed at civil engineers who are taking on management responsibilities for the first time, and has evolved from papers given at a half-day meeting held at the Institution of Civil Engineers in October 1986. Since then, the Authors have updated and expanded the papers they gave on that occasion.

One of the most frequently expressed needs of engineers is for management training, which is not surprising considering that the 1984 ICE Survey showed that the managerial content of the jobs of 78% of the corporate members was equal to or greater than the technical content. The most common difficulties which members experienced were in motivation of people and in industrial relations.

What is meant by 'management'? There are many definitions — getting things done through other people, and co-ordinating the resources of finance, plant, materials and labour in the pursuit of declared objectives, are just two.

Whatever definition is used, employees tend to think that managing is something done by those holding positions at the top of the organization, but this is not so. As soon as an individual has responsibility for even one other person, he or she becomes involved in managing. Such a person therefore should have management knowledge and skill if he is to discharge his responsibilities effectively.

Many of those who attend management courses come away disappointed. They had expected to learn how to manage, but that is not a valid expectation because of the many variables in every situation. However, one can learn from the experience of managers, and from the work of thinkers and writers on management, about the kind of action that succeeds in many situations, and that which fails.

The section head or group leader needs to be more than a supervisor — he or she should be a manager who is responsible for all aspects of the work of the section and the people who comprise it.

To illustrate points of interest and importance, anecdotes have been incorporated into the text and are shown as 'cases'. This practice has also been extended to include quotations.

Contents

selection; induction and training; performance
appraisal; management development systems; self
management; further reading

1 Introduction

Concepts

Learning about managing people presents a challenge to engineers accustomed to dealing with predictable things like mathematics and the properties of materials.

When it comes to managing people, one is dealing with ideas and concepts as well as with the personalities and foibles of individuals, and people are less predictable than, for example, mathematics and materials.

Thus, one of the difficulties of handling people is that they are all different and react differently, even to apparently simple things like receiving instructions. A method which works well with one individual or group of people may not do so with another person or group. Individuals themselves can differ from day-to-day; be affected by domestic, financial or emotional problems. What appears to one person to be a problem is as nothing to another.

Nevertheless, what may seem at first sight to be a vague subject does begin to make sense when one learns what it is that motivates people and that organizations do have recognizable forms — that some styles of managing are more effective than others, and that there are principles of management which can act as a general guide to everyday management.

Style

Some management pundits claim to offer universal panaceas for all management ills, but none of them will be successful except in the short term or in a particular set of circumstances.

At one time 'communication' was offered as the solution to all managers' problems; then it was 'management by objectives';

then 'planning, programming and budgeting systems'. While these are valid techniques which have their place, they are just part of a large army of methods to be deployed as each situation warrants.

Managers cannot succeed by following just one principle, theory or technique, but should select what suits their style and then formulate their own approach. This approach can then be modified depending on the individual characteristics of the people involved and the problem being faced.

So how does one actually manage? What is it that the manager does day-by-day, hour-by-hour?

At one moment he or she will be leading, at another advising, deciding, listening, or doing one of the numerous other things that are needed to keep the team, group, section, department or whole organization on target.

There is no single solution to the challenge of managing — every person and every situation is different. But there are some fundamentals of successfully managing people which the budding manager should learn, and those are the fundamentals with which this book is concerned.

Abilities

Organizations are composed of people of mixed abilities with some capable of more than their position demands and others at, or above, their level of competence. Steps should be taken to allow those who can, to achieve more. Anyone in a position of authority needs to understand the problems of organizations, to take steps to tackle them, and then to share with others the results of his or her experience.

In reading about types of organization there is much that can be recognized from personal experience: the problems of over-formalized systems; attitudes and abilities in other people that one admires; disappointments over employees who are not as committed to achievement as oneself; remoteness from the top; lack of reward for initiative.

A good start can be made by reading suitable books or articles, or hearing talks or lectures about management, but one must question what one reads or hears and consider how it fits in with one's own experience. One has to read, think, observe, discuss, and try things out.

Management takes place around us every day — in the office, on site, in the home, at the club. We should observe it in action and learn what kind of action succeeds and what fails. Although there may be particular aspects of civil engineering and building which place different demands on its managers, it remains true that managing people is essentially the same in all industries. Thus, much useful knowledge can be gleaned from the experience of others in different fields.

Attitudes

It is a paradox that some people will put a minimum of effort into their work, yet a great deal into other activities. But what is work? For example, for some employees gardening is work, but for others it is a relaxing, interesting and absorbing hobby.

It is not so much the content of the activity that distinguishes toil from recreation, as one's attitude to it, the amount of time devoted to it, whether it is paid employment, and perhaps most important of all, whether one has a superior. Mark Twain said 'Work consists of whatever a body is obliged to do, and play consists of whatever a body is not obliged to do'.

Nevertheless, employment gives more than remuneration; it confers status, identity and a sense of purpose. Managers should strive to gain from employees the same enthusiasm and commitment for work as they give freely to their hobbies and outside interests.

Motivation

Chapter 2 deals with performance and motivation. Theories of what it is that motivates people make fascinating reading and Maslow's well-known hierarchy of needs forms a basic framework which everyone can comprehend.

Building on this framework, the work of Herzberg and McClelland is discussed, including the more detailed aspects of motivation which the manager must know about such as power, influence, responsibility, freedom, achievement, rewards and job satisfaction. A key factor in the management of individual and team performance which is covered is the setting of goals.

Poor communication has its effect on performance and motivation and so ways of improving communication are discussed. Furthermore, understanding something about groups and how

3

people can work together to achieve greater performance than the sum of that of the individuals in them is part of the manager's portfolio of skills which is dealt with in this chapter.

Organizations

The framework of organizations is dealt with in Chapter 3. While small organizations have simple structures, larger ones demand a more complicated arrangement. Inevitably this makes them more formal, with rules and procedures that are intended to bring a degree of uniformity to the way in which they operate. Yet there is much that can be done to make a formal organization more responsive to the desires and abilities of employees.

However, it should be remembered that many employees are satisfied with the orderliness, security and lack of uncertainty to be found in large, impersonal organizations.

Various systems of organization, from bureaucracies to organic systems are discussed and it is shown that the demands of different organizations, such as the large local authority department or the small consulting firm, will have a bearing on the system which will be most appropriate.

The chapter illustrates principles of structuring and gives examples of organizational structures in action in a promoter's firm, a consulting engineer's office and a contracting company. These examples demonstrate the complexity which exists in organizations day by day.

Managing

Chapter 4 begins by summarizing those principles of management which Henri Fayol, the French mining engineer and industrialist, prepared over 70 years ago. They have stood the test of time well.

Various aspects of day-to-day management are discussed, including span of control, line and staff roles, relationships with colleagues, and delegation.

The way in which subordinates are perceived by managers is an important aspect in the success or failure of an organization, and the views of Douglas McGregor are briefly set out.

The very important matter of leadership in organizations is also covered, together with the related matter of management style. The chapter includes a discussion on the daily problems of managing and concludes by dealing with supervision.

Development

Personnel development is important enough to have a chapter to itself, and Chapter 5 deals with various aspects of this subject. As well as the overall concept of manpower planning, it also covers job specification and the need to modify job descriptions to suit the abilities of the people employed.

Selection of personnel is a difficult task, made more so by recent legislation, and various techniques designed to assist in choosing the best person from a number of applicants are discussed.

Induction of new recruits and their subsequent training are necessary for the good of the organization yet they are matters which are not always adequately catered for.

Performance appraisal is a knotty subject which, if not properly carried out, can do more harm than good. This, and management development, are also discussed in this chapter. Some basic ground rules for managers are set out. The need for self-development is discussed.

Trade unions

At one time, the spectre of militant trade unionism was something which some managers dreaded. The threat has, at least for the time being, receded. But there is much more to trade unions than militancy, and it behoves the competent manager to know something of trade unions, their aims, and the services they give to their members.

Chapter 6 tells us what trade unions do, and explains their central feature — collective bargaining. The role of the shop stewards, negotiations, conciliation and immunities, as well as the working rule agreement for the construction industry are covered and discussed.

2 Performance and motivation

Introduction
It is very important at the outset to be clear about the meaning of the word 'management' in the context of this chapter.

Management in the construction industry is all about managing performance — the performance of businesses at one end of the spectrum, and individuals at the other. In between these two extremes lie:

- design office groups or teams
- gangs
- sections
- project teams
- sites or design offices
- specialist functions, e.g. estimating, engineering, surveying, buying
- operating divisions of companies, authorities etc.

All across the spectrum, the nature of management is the same:

- good analyses of the problems and opportunities inherent in situations have to be made
- good solutions have to be found
- good plans have to be made to implement the solutions
- the team which has to turn those plans into reality has to be:

  o assembled in the right quantity with the right skills
  o organized
  o briefed
  o monitored for progress against plans

o   helped to adjust to changing circumstances or shortfalls in performance

o   motivated to win.

To do all this well requires skills and qualities ranging from sensitivity, perceptiveness, courage and tenacity, through analytical problem solving and planning skills to communicating and motivating skills. In a nutshell, successful business management needs all the skills and qualities which we associate with leadership.

It is often asserted that 'leaders are born — not made'. It is certainly true that many great leaders were born with a combination of talents and qualities which enabled them to become charismatic leaders, often at very early ages. However, it is also true, fortunately so for the great majority of us, that most people can acquire enough of the skills of leadership to enable them to be very effective managers of the performance of the team for which they are responsible.

Much attention is paid to what are sometimes called the 'hard skills' of management — for example, analysis, decision making, organization design, cost and progress control systems, planning etc. — but less attention is paid to the development of the 'soft skills' of motivating people.

To be effective, managers must possess a blend of the hard and the soft skills. The interaction between these two skills is a fascinating subject. None of us can be highly motivated, working for someone we do not respect. In our profession, respect tends to be based on skill and achievement. Achievement is rare without skill, although luck always has a part to play. Thus, our ability to motivate is a function of what we have achieved and hence our own innate and acquired 'hard' skills. If we are to motivate members of our team, we must have enough of the hard skills to make good decisions and command respect. However, even if we can make good decisions, we will achieve little unless we can motivate the individuals who comprise the team which we are leading. That team may be the board of a multimillion pound construction company, a division of a water authority, or a section of a design office. As managers, the problem is the same for all of us. We have to obtain maximum performance from individuals as members of teams.

Individual performance

The performance of an individual is a function of two things: his (or her) level of ability and motivation. Cases 1 and 2 illustrate this point.

The solutions to the problems described in Cases 1 and 2 are straightforward: train Dave Wren and motivate Oleg Topolski. However, the latter is more difficult. Oleg Topolski's boss was, clearly, not a born leader, or he would have managed the situation in such a way as to keep him motivated. What techniques

---

Case 1.

Dave Wren, a section engineer working on a major bridge site, was very keen to succeed and progress in his career. He worked all the hours God gave, often at home, on plants, temporary works designs etc., but still his section fell behind programme and costs were above budget. Why?

Because setting out errors caused delays and rework costs; bending schedules prepared on-site, in this case, contained errors and omissions; labour output was low as the result of the frustrations this caused.

Motivation in this case was very high, but, alas, the basic skills were deficient. Performance was low as a result.

Case 2.

Oleg Topolski, the chief designer of one of the country's leading specialist contractors, is widely acknowledged to be brilliant and probably the best in his field. However, the company is losing tenders as a result of conservative designs; contract costs are over-running due to design errors and the company's reputation is suffering from the resulting late completions. Why?

Because he has been passed over for promotion. Apparently, to get on in this company you have to be a fast talker, play golf with the Managing Director, and the less you know about engineering the better. Anyway, Oleg is keen on amateur dramatics so is not too concerned that internal politics has robbed him of all his influence on day-to-day engineering matters.

In this case, the Chief Designer has all the necessary technical skills to do his job effectively, but his performance in the job is very poor as a result of his very low level of motivation.

---

are available to him to raise motivation and, as a result, performance?

Motivation theories

An understanding of the theory of motivation can be very helpful to a manager in generating the solutions to a motivational problem or in creating the right conditions for high levels of motivation. In this respect there is an exact equivalence with the usefulness of a knowledge of the theory of structures in solving structural problems or designing effective structures.

However, the similarities between behavioural theories and engineering theories are very short lived. By and large, engineering theories only survive if:

- they can be demonstrated to be true to high levels of certainty after exhaustive testing

- most people accept them.

However, behavioural theories deal with:

- people
- situations
- the interaction between the individual and the situation.

As all of these are infinitely variable, behaviourial theories tend to survive not on the basis of rigorous proof but on the basis of:

- plausibility
- usefulness to practitioners in that solutions developed on the basis of them tend to be successful.

With this in mind, we can go on to examine, rather briefly, a selection of the best known and most useful theories. However, it is worth noting that all of them are, fundamentally, 'What's in it for me?' theories.

As individuals, faced with a demand made on us which requires us to decide how much effort we are prepared to dedicate to responding to that demand, we go through a process, sometimes both instant and subconscious, of asking 'Why should I? What do I get out of it as a reward for my efforts?'

Expectancy theory

This theory examines the thought process that we go through

9

when deciding how much effort to put into a task or, in other words, how motivated we are to do what is being asked of us. The theory asserts that six questions are inwardly addressed.

- 'If I succeed in doing what is being asked of me, what will the pay-off or outcome be for me?'

  By pay-off we may mean anything from keeping our jobs, through enhanced promotion prospects, bonuses etc., to such feelings as pride, self-satisfaction and being true to oneself.

- 'How important to me is the pay-off?'

  If it is very significant to us we are likely to put a lot of effort into achieving it. If of only minor significance then only minor efforts would be forthcoming.

- 'Am I capable of achieving what is being demanded?'

  We may be very enthusiastic about the potential pay-off, but if the chances of winning are negligible, the effort we are prepared to exert will also be negligible. Our chances of success are, of course, determined both by our own abilities and factors external to ourselves. Such questions arise as:

  o 'Will the rest of the organization operate with sufficient efficiency to let me win?'
  o 'Will suppliers or customers perform their obligations?'
  o 'Can I, or will I, get the necessary equipment and resources of the right quality to permit me to win?'

- 'What probability is there that if I succeed then I will get the reward?'

  If the reward is worthwhile and our chances of succeeding are high we will apply the necessary effort. Unless, of course, there is a history of broken effort/reward linkages. This can occur because an organization breaks its promises, managers promise what they cannot, or are not allowed to deliver, or individuals second-guess incorrectly as to what the reward will be when making their decision.

- 'What happened last time I put in a major effort and succeeded?'

  Almost as important is 'Do you get the same rewards regardless of effort and success?' If there is no consistent and valid link between effort, success and rewards, then this will have a major impact on the decision on how much effort to apply.

- 'Am I absolutely clear about what I am being asked to do?'
  If we are not clear about what we are being asked to do and how success will be judged, then it is difficult to be clear about how much effort to put in. If there is a history of tasks being changed, 'goal posts' moving, then we are likely to be equivocal about effort.

Case 3 illustrates how a good manager takes these six questions into account.

Case 3.

As site manager on a major dam site, Brian Wills had a great talent for stretching his section engineers — what he asked was always within their abilities, but at the limit of them. In turn, they had come to learn that if you did a good job for him, Brian was a very fair and considerate boss; if you didn't, then things were not quite so comfortable!

Brian thought that it was very important to ensure that the site was well laid out, that there was a little more equipment present than the plant histograms indicated, and that it was well maintained. He took particular interest in the relationships with suppliers and sub-contractors — ready-mix, crane hire, scaffolders — and with the RE. Clearing the way so that his engineers could perform well was a very important part of his job, he thought.

Communications were very good on-site. Weekly progress and planning meetings and the usual evening 'chewing of the fat' ensured that targets were clearly defined, engineers had a chance to modify the more ridiculous ones and there was seldom a need for drastic change to the short-term plans.

Brian had a highly motivated team. He also had a reputation for slightly overspending his site establishment budget to get things set up properly, but vastly underspending on direct costs as a result of high productivity. People often got promoted off his sites, which was a terrible nuisance but rather flattering to Brian, because he got a kick from successfully developing young engineers and losing one was evidence that other people thought he had succeeded. Because his engineers were rewarded by promotion for their good performance, the required effort was usually forthcoming.

Maslow's hierarchy of needs

This theory addresses the issue of pay-offs and helps us to understand what it is that people value as outcomes, and why they do so.

Maslow asserted that:

- all humans have needs that have to be satisfied

- these needs fit into a hierarchy with self-actualization at the top, descending through esteem, status, ego, belonging and love, security, down to physiological needs

- we attempt to satisfy the lower-order needs first and work up through the hierarchy

- only unsatisfied needs are motivators.

First, as individuals, we have to ensure that we can eat and sleep. We are motivated to do what is necessary to satisfy these needs. In modern society, money is the means by which we achieve this. Having done so, our next goal is security — a state in which we can see our physiological needs being satisfied for the forseeable future. An individual who believes that keeping his job is key to his survival will be motivated by the risk of losing it. An individual who is confident in his ability to get another job if he loses this one has already satisfied his security need, will not be motivated by the risk of losing his job and will need a higher order pay-off to be highly motivated.

Once the lower-level needs are satisfied, man seeks the satisfaction of belonging and affection. The affection of spouse, family, friends and colleagues and belonging to groups, teams, parties, factions, institutions. Once again, money may be the means by which the end is achieved, although there are many ways of satisfying the affection and belonging needs which have nothing to do with money. An unsatisfied belonging need can be a powerful motivator. Anyone who has been excluded from a group, membership of which would be highly valued for whatever reason, knows the forces which are at work and the efforts that will be made to achieve membership.

Next in the hierarchy are our esteem needs, by which is meant the need to be thought well of by others, for what we are or what we have achieved. Yet again, money plays a part because money represents a means by which an individual can tell the world of his achievements. The giving of status by means of job title, quality of car etc., is also an important way of satisfying this need.

Finally, at the top of the hierarchy come self-actualization needs. This is a hard concept to explain but is perhaps most readily understood as personal fulfilment and the satisfaction one gets from solving difficult problems, remaining true to one's principles, and being of service to one's fellow man. Money may play a part in satisfying this need also but, generally speaking, the individual will be prepared to sacrifice money in order to satisfy this need, providing that he already has enough money to satisfy the lower-order needs.

Maslow's theory does much to help explain human behaviour, to explain why money and status are such powerful motivators and to explain why engineers, who have opportunities for self-actualization in their work far beyond other occupations, are prepared to work so hard for such relatively low financial rewards. It also explains why people who have little opportunity for self-actualization at work, drive so hard for higher salaries and the esteem and status which they bring, whether in the City or on the production lines.

Finally, it helps managers to select motivational strategies which are appropriate to the individuals they are trying to motivate. As individuals, we are all motivated by different things. A pay-off which, for one person, is very important and valuable is of no consequence to another. Thus, if we are to be successful managers, we must take the trouble to understand our subordinates, so that we can present to them a set of meaningful pay-offs for good performance.

Herzberg's two factor theory

Herzberg's theory can be seen as an extension of Maslow's, but it introduces the concept of potential demotivators which can exist in the organizational environment. These include such issues as:

- working conditions
- policies – salary, employment conditions, etc.
- administrative efficiency
- style of supervision
- relationships between employees.

These have been termed the 'hygiene factors'.

Herzberg asserts that getting these factors right creates the conditions in which a set of personal needs — what he called the 'motivators' — can produce high levels of effort. However, if the hygiene factors are wrong then even activating the motivators will not be effective.

The motivators:

- achievement
- recognition
- responsibility and freedom
- advancement

are somewhat akin to Maslow's 'needs', while the hygiene factors reflect the issues raised in the third question in the section *Expectancy theory* (p. 10). As can be seen, there is a degree of convergence between these theories, which is reassuring.

However, Herzberg ensures that, as managers, we do not overlook the need to create the necessary conditions in the work environment (efficiency, order, fairness in salary and personnel policy, warmth, quiet, shelter from weather and so on, almost without end) if high levels of motivation are to be reached.

McLelland

McLelland's theory maintains that individuals are motivated by three needs:

- power (or influence)
- affiliation
- achievement.

Each of these needs carries a different weight with each individual; one person may have high needs to achieve and exercise influence, and a low need for relationships, while another may have a high need to achieve, but have no desire for power or relationships.

High motivation will result from situations where an individual has the opportunity to satisfy his psychological needs.

- Power — consider the motivation of those who aspire to be Members of Parliament and Government. There were few more highly motivated people on large sites in the 1970s than shop stewards who held, and enjoyed, wielding enormous power over their members, management and customers.

- Affiliation — consider the motivations of the caring professions.

- Achievement — consider the efforts drawn out by the drive to achieve on time, within budget; to earn a University degree; to run a marathon.

Once again, money can be seen to play a role as a means of exercising power, or as a means of measuring and demonstrating achievement.

McLelland's theory helps get over the hierarchical nature of Maslow's theory because it allows all types of psychological needs to co-exist at the same time and recognizes that self-actualization needs can be stronger than survival needs.

Money

It is often said that the only real motivator is money. However, research has shown that less than 10% of engineers put money at the top of their list of motivators (Landis, 1968). The fact that construction is a relatively low-paid profession, yet there has only recently been difficulty in recruiting at the university level, tends to support the research.

However, having reviewed the motivation theories, we can see that money can be the means whereby the psychological needs of the individual can be satisfied:

- money, to buy possessions which satisfy belonging and esteem needs
- money, to make it possible to do things which satisfy self-fulfillment needs.

Money is not the only means by which these needs can be satisfied. Job design, job content, and non-money reward systems have persuaded many people to compromise over salary levels and remain very highly motivated.

However, wide use of bonus schemes is made in the construction industry in the belief that operatives will produce more if there is a financial incentive to do so. Opinions vary greatly within the industry as to the effectiveness of such schemes, and there is no unanimity on the reasons for the effectiveness of those schemes which produce good results. Many people believe that success derives more from the following, which are to be found in all successful bonus schemes, than from the money rewards which result:

- the clear goals which are set
- the participation in the setting of the goals which occurs during the negotiation of the bonus rates
- the communication of the performance levels expected
- the high degree of job analysis and organization which is applied
- the recognition which management is giving to the workforce and the interest it is taking in production and productivity levels.

None of the research which has been carried out has provided any conclusive evidence one way or the other.

There are two principal types of scheme which relate pay to production. The first involves completing a set amount of work for a set amount of pay. For example, a kerb layer can earn bonus in direct proportion to the number of kerbs laid. A variant of this type is the 'job and finish' where a task is set and the operative is free to leave when the task is complete, collecting a full day's pay. In this case, the reward for extra effort is free time, rather than money. The second scheme involves setting standard times for executing tasks and paying bonuses related to the amount by which actual output achieved in a given time exceeds the output which would have been achieved at standard rates.

The detailed design of bonus schemes is highly technical and an in-depth examination lies beyond the scope of a book such as this. However, if a bonus scheme is to be effective in raising levels of performance above what they would otherwise be, it will depend on:

- analysis of the tasks in sufficient depth to enable management to set demanding but attainable levels of output

- negotiation of a bonus structure which motivates those concerned to strive to achieve the high levels of output sought

- accurate and objective measures of output which provide a minimum of scope for disagreement and argument

- management's ability to organize the availability of work areas, access, materials, plant, equipment and so on in such a way as to allow work to proceed at a rate determined by the workforce as opposed to management
- strong co-ordination between different trades on different bonus schemes opting to work at different rates
- clear understandings on who it is that 'pays' where the workforce is prevented from achieving its preferred levels of output by circumstances outside its own control
- thorough quality control and mechanisms for adjusting bonus payments when defective workmanship comes to light a long time after the work was carried out.

Throughout the 1970s, bonus schemes were used extensively and, because they are so difficult to design effectively and agree by negotiation with the workforce, and because conditions on construction sites are so uncertain for so many reasons, experience with them was largely unsatisfactory. During that period, project performance in terms of time and cost was bad.

The trend in the later 1970s towards labour-only subcontracting was, it is thought, very much a response by construction management to the problems of production bonus schemes, enabling them to obtain a fixed price for a fixed amount of work. They leave it to the relatively small and non-unionized subcontractors to reap the rewards of high productivity, paying top rates (not necessarily geared to output) to good workers and not tolerating poor ones. Bonus, if any, is in the form of profit share, but the motivation to perform well comes from demanding goals which are clearly defined and communicated, good organization, and the knowledge that good performance leads to the continuing opportunity to earn good money.

Job satisfaction

It is often said that job satisfaction is the biggest motivator of all. In reality, job satisfaction is the *consequence* of an individual doing a job which is satisfying his (or her) psychological needs (whether they be the ones which Maslow, Herzberg or McLelland have identified, or not) rather than a *cause* of motivation.

High motivation and job satisfaction often go together, but by no means always. A draughtsman who is also a keen dinghy sailor may be very satisfied with his job, because it is challenging, creative, sociable and provides enough money and free time for him to indulge his interest in sailing. However, his motivation may be low because he has no need to earn more and his mind is on sailing, not detailing. Thus it can be seen that satisfaction with a job does not necessarily lead to high motivation.

Lack of job satisfaction is evidence that his (or her) needs are not being satisfied and low motivation is also likely to be found in such circumstances.

Thus, job satisfaction is not something which a good manager tries to provide for his team for the sake of providing it. High motivation is the goal; job satisfaction is a result of it, not necessarily a cause of it.

Organizational climate and leadership style

It would be wrong to conclude a section on motivation theories without making reference to the effect that the culture, or climate, of an organization and the leadership style of one's superiors, has on motivation.

To talk about the climate within an organization is to attempt to describe what it feels like to work in that organization. When the climate is good, we feel good and we feel like making a big effort. When the climate is bad we feel bad. In the latter case, the effort tends not to be forthcoming because we find it hard to answer the question 'where is the organization going; what influence will the achievement of my task have on the overall scheme of things?'

The climate, or culture, of an organization is a complex blend of its:

- management style
- method of supervising and controlling
- view of what is acceptable and what is not acceptable
- reward system
- team spirit
- clarity of purpose and method.

There are echoes of Expectancy theory, and of Maslow, McLelland, and Herzberg's hygiene factors in this. The effect of climate on motivation is very great. Managers therefore need to put the creation of a highly motivational climate high on their list

Case 4.

Morale had been very low in the council's Chief Engineer's office. The Chief Engineer himself had recently retired and there was no doubt that during the last few years this had affected budget allocutions and the Department had lost its drive and sence of direction.

The office itself had become tatty, the internal mail system was very unreliable and its filing system had fallen into such a state that one really had to keep a personal copy of every item.

There was a lot of internal feuding between sections; back-covering and point scoring. Frankly, it mattered very little how well one did one's job because promotion followed on a 'Muggins turn next' principle and when something went badly wrong nobody seemed to worry. When something was done really well it was taken for granted.

There was a very high level of criticism in the local press, much of it unjustified and unfair, but not all by any means. No attempts were made to put the record straight.

Of course, in the public sector, salaries are highly structured, so there was no real chance to benefit financially from making a major effort to do a good job.

When the new Chief Engineer joined from another authority, he was quick to sense the low morale and analysed its sources as:

- low standards of performance
- no rewards for good performance, no penalties for bad
- individuals did not feel responsible and were not given the responsibility for achieving results
- there were no feelings of team spirit because the mission of the department had not been recently reviewed and re-stated
- neither the way in which the department's performance was to be judged, nor the levels of performance to be achieved had been formulated or communicated
- the senior managers had not, together, determined how to reach those levels.

Although he had to accept that a large integrated department with public accountability had to be somewhat bureaucratic, at least the procedures should be clear and efficient and should take advantage of the latest computer technology.

He could not motivate by encouraging staff to take risks, but he could still set some challenging performance standards which would stretch them to the limit. Promotions would only go to high performers. He would cultivate relations with the local press and would make sure that they knew about all the good things that his department were doing and were going to do.

He knew it would take time, but he was confident, given the calibre of people in the department, that he could build them into a team that was highly motivated and proud of what they could achieve.

of priorities so that, within that climate, they can work on motivating individuals by helping them to answer the question 'Why should I?'. Case 4 should help to illustrate this concept.

Self-starters

There are some people who, although badly managed and operating in a demotivating climate, nevertheless exhibit high levels of motivation and dedication. This can be explained in terms of the rewards they reap from good performance in such circumstances, mostly of the self-fulfilment and esteem type.

The self-starter helps us to identify that:

- different people have different perceptions of the environment or climate necessary for high motivation to be developed

- the level of motivation developed is a function of all the factors. If all are strongly supportive of high motivation, then motivation will be higher than if some are not. It does not follow that if some are very low then motivation will be very low — it will merely be lower than could be achieved.

Summary

Figure 1 shows how all the elements of the motivation theories presented in this section interlink and interact.

Objectives

It follows from the discussion of motivation theories that a key factor in the management of individual and team performance is the setting of goals. When we know what we have to achieve, we can determine the level of effort to apply.

If a manager is trying to achieve high performance, or lift performance, than all the theories tell us that the goals he sets must be:

- demanding, so that performance is maximized and personal satisfaction can be obtained from achieving them

- achievable, so that the individual is not demotivated by being set impossible targets and getting no pay-offs

- measurable, so that they can be precisely described and the individual can know, objectively, whether he has achieved

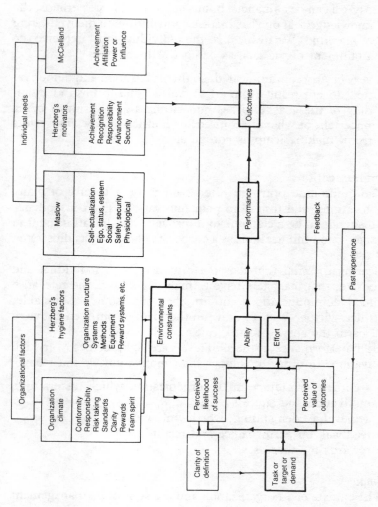

Fig. 1. Interlinking and interaction of motivation theories

them or not and therefore has the right to 'collect' the reward in whatever form it may come

- worthwhile, in terms of the pay-off for the individual. The pay-off can be a financial bonus at its crudest or, perhaps, the knowledge that one has made a contribution to the well-being of mankind. Whatever it is, the individual must perceive the attainment of the goal as being worthwhile

- very clearly communicated, so that frustration and demotivation do not result from effort wasted due to misunderstanding of the goal, effort is not withheld as a result of being uncertain of what is required, or inefficient working results from 'making it up as you go along'.

Communications

Bad communications are often seen to be a feature of many organizations, and to lead to poor motivation. The consequence of this tends to be a clamour to improve communications and to hold good communications as a corporate objective, as illustrated in Case 5.

Bearing in mind that it is performance, both individual and corporate, that managers have to manage, what is the relevance of good communications, or are they just some fashionable, worthless virtue? We can draw on the theories outlined above to lead us to the answer to this question.

High performance, if the skill levels are right, comes from high motivation. In order to achieve high motivation we must have:

- a climate or culture in the organization which fosters high motivation and success
- achievable, demanding, worthwhile goals
- feedback on performance and results
- rewards for success.

Climate

The climate of an organization and the style of its management are key factors in motivation. They spring from a vision of the future and a clear view in the mind of the Chief Executive of what he (or she) wants the organization to be.

In order to create the climate and achieve the Chief Executive's

Case 5.

The parent company of a construction company was concerned about morale in its subsidiary and commissioned a survey of the state of communications within it. The results were very worrying. Communications were thought by the vast majority below the subsidiary Board level to be very bad. The Board thought communications were very good!

As a result of the survey, and in deference to their parent company, the Board adopted as one of their key corporate objectives 'to improve communications'. Five years later a similar survey produced almost identical results. Why?

Because the Board knew instinctively and correctly that improving communications would win no more tenders, complete no more contracts, reduce no costs, and add nothing to the bottom line. 'You cannot take good communications to the bank or use them to pay suppliers', said one director. They had not acknowledged that to improve communications was a goal worth achieving, had made little effort to improve, and so little had changed. Had they believed that to improve communications would result in higher profits they would have put much more effort into it.

What they had failed to appreciate was the linkage between good communications and motivation and the linkage between high motivation and tender conversion ratios, contract costs and profit.

goals, this vision has to be communicated to every individual in the organization, who has to understand, through a process of informing, questioning, persuading and reassuring, where and what the organization is trying to be. To employ an outdated naval analogy, the stokers stoke with much more enthusiasm if they know where they are going, and want to get there, than if they don't.

In order to create the right climate, the organization must not only communicate the vision, but must also communicate to its staff, examples where individuals have succeeded in matching its expectations and examples where expectations have not been met. One supports the development of the climate, the other keeps it developing in the right direction.

Goals

This has been thoroughly covered above. However, it is worth adding that goals set by individuals themselves, given the right

23

climate, are often more demanding than their manager would dare set. In addition, there is often a need for a manager to convince a subordinate of the worth of a goal — why it is needed, what benefit it will be to the common good, how it can be achieved.

Thus, communication on the matter of goals is vital to:

- achieve clarity of purpose
- take the individuals' views into account
- establish that the goals are worthwhile
- establish achievability.

Feedback

Teams and individuals have a need to know how they are doing. It encourages them, and therefore motivates them, to hear that they are progressing well towards the achievement of their goals. If progress is not good then there is a problem to solve. This is nearly always best done by bringing the problem into the open and getting the team or individual to participate in finding the solution.

In all cases, feedback demands communication and is a very important part of the motivation and performance management process. Not only does the individual need to know how he is doing in his job of achieving his goals, but he also needs to know how the organization, of which he is part and to whose vision he has committed himself, is getting on with the achievement of its goals.

Thus, feedback communication must take place on many levels — corporate, divisional, departmental, team and individual.

Rewards

We have seen how important the personal pay-off is to motivation. In the section on expectancy theory, above, we saw how a lack of clear understanding of the pay-off could lead to disappointment on successful achievement of a goal and thus to a downward spiral of motivation or a holding back of effort. Thus, to be clear about pay-offs is fundamental to high motivation. The communication of them is, therefore, vital.

A problem arises for a manager if:

- he is not clear himself what the pay-offs can be, or

- the pay-offs are not very attractive.

If this is the case then there is a climate problem in the organization and only three possible courses of action exist:

- change the climate
- accept low motivation
- use the motivation theories to find some worthwhile pay-offs which can be delivered.

Recognition is a very effective motivator, as all the theorists agree. If a climate has been created which recognizes achievement, then achievements have to be communicated to the rest of the organization, both to support the development of the climate and to provide personal rewards, albeit in the form of recognition as opposed to more tangible benefits.

For example, the Queen's Birthday Honours system is a very effective motivator for those who are motivated by the need for status and esteem. If nobody knew what the system was, the Honours List were never published and titles were not used, it would be completely ineffective. Thus, communication is essential to the success of such reward systems.

Mechanism

For a wide range of reasons, therefore, it can be seen that good communications, while not an end in themselves, are a very important means to the end of achieving high motivation.

There are a number of ways of achieving effective communications and they fall into two main categories:

- Formal
- house journals
- employee annual reports, oral and written
- training courses
- team briefings
- written policies and procedures
- performance planning and appraisal
- company award functions

- Informal

- job related discussion involving horizontal and vertical sections through the organization

25

o social occasions which mix senior and junior across functions and departments
o senior staff using job related opportunities.

Good communications will be composed of a mix of all the methods without being overly dependent on any one. It is common to find an over-dependence on the informal, so that all information is carried on a 'grapevine' which cannot be controlled or directed to align with the corporate goals. However, over-dependence on formal systems will inhibit the development of a healthy climate, which depends largely on senior management's enthusiasm for their visions 'rubbing-off' onto their staff. This is difficult to achieve without face-to-face informal communication taking place.

A common mistake is to use each channel of communication for many purposes. For example, house journals are sometimes used simultaneously:

- for public relations purposes; sent to clients, competitors, job applicants, etc.
- for rewarding individuals by recognition
- for building role models
- for climate building
- for informing staff of interesting news – orders received, investments in new equipment, etc.
- to satisfy belonging needs – marriages, births, staff appointments.

All these are valid for a house journal, but some of them conflict and others would be more effectively achieved by other methods.

Similarly, if the informal system has to carry all the information about corporate strategy, short-term objectives, career progressions, conditions of employment, who is succeeding and who is not, both misunderstandings and knowledge gaps will result.

Senior management has to decide:

- what has to be communicated
- why
- to whom
- which is the most effective vehicle

and then avoid overloading or misusing the various vehicles which emerge from that analysis.

Groups
Introduction

We have discussed the management of individuals and the issues which have to be addressed if, as managers, we are to get the best out of our people. However, most managers are confronted with the challenge of getting the best out of teams of people, rather than from individuals as individuals.

It is intuitively obvious to many people, but has also been dramatically demonstrated in many ways by many researchers, that the performance potential of groups is much higher than that of individuals. This is no more than an academic restatement of the old adage that 'two heads are better than one'.

It is, therefore, an important part of a manager's portfolio of skills to have the ability to make the performance of the team greater than the sum of the performances of the individuals who go to make it up. As we shall see, and as most people have experienced, it is all too easy to get less out of a group than out of individuals, as a result of poor management.

The great gap in performance between how well and how badly a team can perform in a business situation is largely a matter of energy and how much of the total energy of the team is dissipated internally, and how much is released externally to complete tasks and solve 'real world' problems.

Determinants of performance

How, then, do we manage the performance of teams, given that we know how to manage the performance of individuals?

There are four principal determinants of the performance of a team:

- its composition
- the strategies it adopts to achieve its purpose
- the motivation of the individual team members
- the effectiveness of the working relationships between team members.

Composition

This covers:

- the blend of skills and the level of skills.

Given the nature of the task in hand, we must have in the team all the disciplines we need. For a water supply project, for example, this could include: socio-economists; hydrologists; civil, chemical, mechanical, electrical, and control and instrumentation engineers; lawyers; and financiers. If we are to be very successful, each of them must be very good at their particular speciality. Picking the team is therefore very important.

- The number of people in the team and how they are structured.

The team must be big enough to tackle the task in the time required. Giving a major job to a small team may be asking them to achieve the impossible.

Of course, the larger the team the more difficult it is to maximize the performance of it; and structural issues have to be addressed as set out in Chapter 3.

- The blend of the roles which team members are capable of playing.

To be really effective all teams need:

- initiators — to propose solutions, have ideas, think of new ways of doing things

- questioners — to clarify, challenge and test information, assumptions and proposals

- visionaries — to identify objectives and take overviews of strategies, problems and solutions

- problem solvers — to worry about the details and come up with answers

- evaluators — to keep an eye on the progress being made and the quality of what is being achieved

- leaders — to keep the group working well together, lifting morale when it is low, solving problems when the group is in difficulty, creating team spirit, setting standards, resolving conflicts and so on.

Different roles may be performed by different people at different times. One person may fulfil more than one role, but it is essential that all roles are covered when picking the team.

Imagine a team with only initiators and visionaries — nothing would ever get finished. Imagine a team with only questioners, problem solvers and evaluators — nothing would ever get started. Balance is what is required for effective team work.

Strategies

A good team from the point of view of composition, motivation and relationships can still perform poorly as a result of having chosen the wrong strategy. Case 6 gives an example of this. Just like individuals, teams can select the solution, from a number of alternatives, which turns out to have been the wrong one. This is less likely to happen if the team is well picked, is highly motivated and works well together, than if none of these things is true, but it still happens.

Motivation of individuals

Clearly, to perform at its best a team requires each member to apply maximum effort. However, it has to be recognized that the team, its composition and methods of working, have an effect on the motivation of its members.

Case 6.

A process plant contractor had earned himself very unwelcome bad publicity, and removal from one client's bid list, as the result of a problem with one of its installations.

The Operations Director was particularly annoyed with himself because it could have been avoided if only they had decided, a year earlier, to adopt the solution which eventually cured the problem. He had done all the right things and still lost!

It had been a difficult case of a very advanced, highly automated process just failing to meet its performance specification. When first discovered, the problem had been tackled by the top specialists in the company, from all relevant disciplines, together with two top academics. Between them, after hours of discussions, they had unanimously rejected plan B in favour of plan A.

During a year of very hard work, this good team of people put plan A into effect with great skill, earning nothing but criticism from their client for the still out-of-specification results.

In the end plan B was adopted as the only alternative and it worked, but by this time both credibility and money had been lost and it would be a long haul to recover both.

For example, rivalries within the team can be very demotivating for both protagonists and their colleagues. Low standards of work from colleagues, poor communications, failure to have his point heard and recognized can all cause a team member to withdraw his effort. These kinds of problems have both a direct effect on the output of the team by virtue of mistakes and poor decisions, and an indirect effect by reducing the effort than some team members are prepared to put in.

Working relationships
It feels good to work in an effective team where all the members are working well together. To create this good feeling there are two types of issues which have to be addressed:

- climate issues
- process issues.

Climate.   In this context climate can be viewed in very much the same way as in the section on organizational climate and leadership style, above.

- Responsibility and risk taking — are team members allowed to get on with their bit of the job relatively unimpeded by constant need to refer back? Are they trusted to do the right thing, or are they closely controlled?

- Clarity — is it clear to everyone in the team what the objective is, how and when it will be achieved, who is doing what, by what means, with what information, from whom? Clarity is vital to motivation and performance, and demands excellent communications between team members.

- Standards — are the standards of behaviour and performance which the team has adopted acceptable to the individual member?

- Team spirit — do team members trust each other and enjoy working together? Do they support one another when one is struggling or do they blame, criticize and gloat? Can they resolve differences of opinion? Effective team working requires high levels of inter-personal skills.

- Rewards — does the reward system within the team encourage compliance with the agreed standards and promotion of team spirit? Does it discourage, or even punish, failure to live up to expectations? Peer group reward systems can be very powerful motivators.

Processes. Other influences on both the motivation of individuals and the effectiveness of teams come under the general heading of processes.

- Decision making – how are decisions made? Collectively, reaching consensus, or autocratically? Are individuals part of the process or are they told what the decision is, with or without attempt at persuasion?

- Task execution – what methods are used for carrying out the packages of work which combine to make up the whole? Are they modern? Efficient? Appropriate?

- Information exchange – what methods are used for identifying, generating and exchanging the information needed to complete the task? Planning techniques? Document flow control? Co-ordination and progress meetings? What methods are used for giving the team feedback on its performance and agreeing new, challenging, worthwhile goals? How effectively are meetings chaired?

Conclusion

Figure 2 shows how all these influences on the performance of groups interact. An understanding of them enables a manager responsible for achieving results through a team of people, to develop strategies for high performance.

Changing conditions

In the lifetime of an organization, it will go through many phases: success; failure; growth; decline; change. Each condition poses different motivational problems. It requires different organizational climates and structures, permits or constrains a motivationally supportive operating environment, influences the types of goals which are set, demands which are made and, hence, the pay-offs which are available.

In finding a path through such turbulence, business leaders will only succeed if they can carry with them a highly motivated

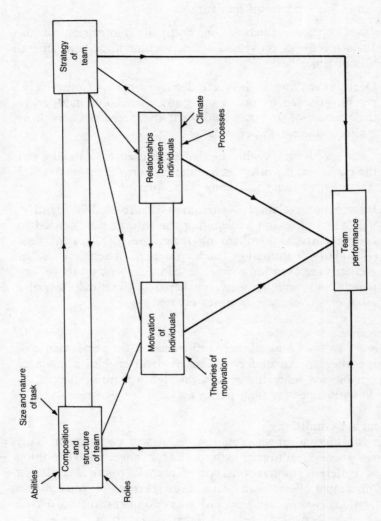

Fig. 2. How influences on the performance of groups interact

team. Those high levels of motivation will come from:

- a well communicated vision of the future and a supportive organizational climate

- creative use of motivational theory to lift levels of performance by the development and clear communication of:

  o demanding but achievable goals
  o satisfying job design and reward structures
  o performance review and feedback systems

- creation of efficient operating conditions and systems

- development of the skills of the organization to match the task of achieving its goals.

Further reading

Adair, J. *Effective leadership.* Gower, London, 1983

Handy, C. *Understanding organizations.* Penguin, London, 1987

Landis, *What makes technical men happy and productive.* Research Management. Industrial Research Institute, Lancaster, Pennsylvania, 1968

Nadler, D. A., Hackman, J. R. & Lawler, E. E. *Managing organizational behaviour.* Little Brown & Co., Boston, 1979

3 Organizations

Introduction

An organization is a group of people working together in the pursuit of common aims.

This chapter considers the framework of organizations and, among other things, it deals with organization theory, the need for structure, the effects of scale, and principles of structuring, as well as giving examples of types of organization structures found in civil engineering.

A large amount of research has been carried out into organizations and many management writers and thinkers have tried to discern what it is that makes some types of organizations more successful than others. Included in this chapter are précis of the more important organization theories that have validity in the construction industry.

Organization theory is defined by D. S. Pugh as the study of the structure and functioning of organizations and the behaviour of groups and individuals within them. As may be expected, work on organizations has moved from a fairly narrow concept to a broad consideration of all aspects of work.

First, there was the scientific management approach pioneered by F. W. Taylor around the turn of the century, which concentrated on shop floor organization in manufacturing. Then came Max Weber's three basic types of organization.

The problems of industrial production during the First World War began to interest behavioural scientists such as Mary Parker Follett, who studied the problems of personal adjustment experienced by people entering industry for the first time. The behavioural scientists became interested in the environmental aspects of work such as light, ventilation, work breaks, and the length of the working day, all of which they studied in detail.

The classical school of management theorists, who were mostly practising managers like Henri Fayol (*see* Chapter 4) sought to develop principles of management which would be applicable to all organizations.

Hawthorne Works

A landmark in these studies was the Hawthorne experiment conducted by Elton Mayo over several years at the Western Electric Company's works in the 1930s. During this period, numerous changes were made to working conditions, including some to working hours, rates of pay, and rest breaks.

Output rose following most of the changes, even when these involved a return to the original working conditions. It was concluded that the cause of the increased efficiency was the marked improvement in the attitude of operators and in their morale, which derived from the development of personal friendships among the operatives and their freedom from supervisory pressures.

Thereafter the wider organizational context in which people worked was regarded as a social system and the results gave rise to the human relations school of management thought.

Although work on organization theory has continued, it cannot be said that there is one all-embracing theory which is acceptable to everyone in all circumstances. Each profession, organization and individual will have its (or his/her) own preferences, depending on knowledge, experience, attributes and personalities.

The need

The need to form organizations to get things done is not new. History gives many examples of early national and city organizations engaged in public works, including irrigation, roads, bridges, the building of monuments such as the pyramids, and the pursuit of conquest.

Aspects of organization are referred to when Moses is told by his father-in-law to seek out capable, honest and incorruptible men, and appoint them as officers of units of 1000, 100, 50, and 10 men (Exodus 18:17, et seq.).

In his book *Anabasis*, Xenophon covers many aspects of organization and management in his description of the long

---

Case 7.

First a firm's dependence upon a market demand must be analysed. Then the sequence of activities in carrying out the primary task can be defined and the working links between activities specified, so providing the logical basis for creating or changing the internal system.
S. H. Wearne, *Principles of Engineering Organization.*

The primary task or set of tasks is that which an organization must perform to survive; the system of organization is the means of achieving it.

---

march of 10 000 mercenary troops from Babylon back to Greece after the failure of their campaign in 403 BC. He deals with leadership in some detail and shows how leaders were selected by popular choice, and frequently mentions consultation with subordinates having taken place before decisions were made.

An organization brings together people with different skills to contribute their abilities, ideas and labour to common goals; it increases the power of an equivalent number of individuals through co-ordination; it can give economies of scale; it enables resources to be used to the best advantage in improving quality and productivity; and it fulfils people's social needs.

The size of an organization affects the number of levels within it, the need for duties and responsibilities to be defined, and the extent to which a formal structure is required. Most organizations tend to have a life span which is greater than that of its members, who come and go, each making his or her individual contribution.

The success of an organization depends on the quality of the individuals working in it as well as on the suitability of its form or structure and management style. Thus, a choice of structure cannot itself ensure that people are well utilized and enjoy their work, but an unsuitable structure will affect performance. A structure should therefore be chosen carefully and be adapted as changing circumstances require.

Organization theories
Weber
Many management thinkers and writers have propounded theories of organization. Among the foremost of these was Max

---

Case 8.

A specialist firm dealing with pre-stressed concrete provided a design service to the employer from an early stage in the development of a project, up to the contract for the construction being awarded to the general contractor.

Under this contract, the specialist firm became a nominated subcontractor to the main contractor to carry out the pre-stressing of the structure which it had helped to design.

There were delays, and the main contractor, with some justification, blamed the subcontractor. The root cause was that the subcontractor had failed to appreciate that his primary task as a subcontractor was to provide the necessary resources, both technical and physical, to the main contractor to enable construction to proceed to an agreed programme.

Instead he had continued in his role as a designer, and had accommodated design changes made by the employer without regard to the effect of his work on-site as a nominated subcontractor.

Weber, a member of the classical school who classified organizations into three basic types depending on the basis of each one's legitimate authority:

- *On rational grounds* — that is, legal authority giving the right of those in authority to issue instructions, e. g. a bureaucracy.

- *On traditional grounds* — that is, deriving from an established belief in the sanctity of immemorial traditions, and the legitimacy of those exercising authority under them, e. g. a family firm.

- *On charismatic grounds* — that is, deriving from devotion to the specific and exceptional sanctity, heroism or exemplary character of an individual, e. g. the church or military organization.

Weber held that employees in a bureaucracy should be appointed and function under the following criteria:

- they are personally free and subject to authority only with respect to their impersonal official obligations

- they are organized in a clearly defined hierarchy of posts

- each post has a clearly defined sphere of competence

- each post is filled by a free contractual relationship

- candidates are selected and appointed (not elected) on the basis of technical qualifications tested by examination

- employees are paid by salary and are usually entitled to pensions. Salary scales are graded according to the position of posts in the hierarchy

- the post is the sole, or at least the main, occupation of the incumbent

- the employment constitutes a career. Promotion may be by seniority or achievement

- each employee is independent of the ownership of the means of administration

- each employee is subject to strict and systematic discipline and control in the conduct of the office.

Although the term 'bureaucratic' is often used in a derogatory way to describe a system which is slow moving and bound by red tape, Weber considered a bureaucracy to be the most efficient and stable type of organization.

Aspects of the formal model of organization (bureaucracy) are present in many organizations, large and small, but it has limitations, and many attempts have been made to evolve other ways of looking at organizations which are more in keeping with what makes them function. Whether one considers a particular model more, or less, satisfactory than any other model will depend on a number of factors, including one's own experience and attitudes to people and work.

March and Simon

March and Simon (1958) define three broad categories of organizational theory based on assumptions about human behaviour in organizations.

In the first, organization members (particularly employees) are passive instruments capable of performing work and accepting directions, but not initiating action or exerting influence in any significant way.

In the second, members bring attitudes, values and goals to the organization: they have to be motivated or induced to participate in the system of organization behaviour; there is incomplete parallelism between their personal goals and organization goals, and actual or potential goal conflicts make power phenomena, attitudes, and morale centrally important in the explanation of organizational behaviour.

In the third category, organization members are decision makers and problem solvers, and perception and thought processes are central to the explanation of behaviour.

Systems of organization

The problems of making work organizations function effectively have led to many investigations into structures and other aspects of both large and small organizations. After examining problems in the electronics industry, Tom Burns suggested two 'ideal types' of work organization. The one he called 'mechanistic' follows generally the formal, bureaucratic type in which tasks are broken down into specialisms. The second, he called 'organic', suitable for conditions of continual change.

Mechanistic systems are relevant to stable conditions; duties and powers are defined and allocated to posts. Relationships are vertical — from superior to subordinates, and instructions flow up and down between them. Such organizations provide an ordered world in which to work. At lower levels, work is comparatively simple and decisions required are few. At higher levels, there is more freedom of action and involvement in decision-making.

Organic systems are suitable for unstable conditions where problems are new and unfamiliar and cannot be broken down and allocated to specialists. Interaction between employees is as much lateral as vertical, and even communication vertically is on a more equal basis since the superior may know no more about solving a particular problem than the subordinate. Each person is expected to be fully committed to undertaking any task which emerges.

Organizing for problem solving

With his colleague G. M. Stalker, Burns identified three responses to new and unfamiliar problems which face a mechanistic system. When a problem occurs that is unfamiliar or outside a person's own specialism, it is normal to refer it elsewhere, usually upwards. When many such instances occur, blockages can take place if decisions are not made, and people become overloaded with work. The usual advice in such cases is to delegate responsibility and decision-making.

Some problems are solved by creating special intermediaries or interpreters, such as methods engineers. A new job, or possibly a completely new department, may be created. Its survival then depends on the perpetuation of the difficulty through which it was created in the first place. This solution sometimes removes responsibility from where it should lie, and creates more self-perpetuating sections.

Other problems are tackled by setting up a committee, which can function in a mechanistic organization without upsetting the balance of power between individuals or sections. Powerful forces are created within mechanistic systems by interlocking systems of commitments to sectional interests and individual interests.

Systems within organizations

Charles Handy (1987) identifies four types of system as a means of understanding the role which systems play in activating parts of the organization:

- *Adaptive* – those systems which are concerned with fitting the organization into its environment, with shaping its future, dealing with divergences and deciding its policies.

- *Operating* – those systems which are concerned with the daily existence of the organization; the intake, processing and export of its materials or orders or cases, the basic essential logistics of the process of work. They include the operation systems of sales and finance as well as those of production.

- *Maintenance* – those systems which work to keep the organization in a healthy and effective condition; the reward and control systems for people, the linking mechanisms between sections of the organization and differing types of systems.

- *Information* – the nerves of the organization, without which none of the systems would function; this serves the three above, running through them and around them.

Coalitions

Handy argues that organizations — even simple rudimentary business organizations, are very complex. Organizations are more realistically regarded as coalitions of groups responding to a variety of pressures.

Individuals are not always lazy and economically motivated. Authority is not the only basis for influence, nor is economic resource power the only kind of power that matters.

Things are not so rational nor so objective as economic management theories would have them be, not even the apparently 'hard' data of accounts.

It is to be expected that the apparently logical system will not always work out so well in practice; budgets may be ignored or abused, information may become a weapon in the hands of some of its possessors, systems can be used to invade territory and be resented for this reason.

Groups

Groups have been dealt with in Chapter 2 and are mentioned here as a matter to be considered as part of organization philosophy or structure.

Most people like working in groups; their behaviour is influenced by other members and they respond to the collective value judgments of the group, being apprehensive of the isolation that may be the result of not conforming to group standards.

Generally, members prefer co-operation to conflict and will adjust their work effort accordingly, as they are more uneasy about incurring the displeasure of the group than of their superiors.

The importance of the group in devising a structure for an organization has led some people to advocate building the structure round the group, including Rensis Likert and his *System 4* (*see* Chapter 4). His reason for doing so is the superiority that he believes groups have over individuals in both decision-making and supervision.

Demands of size and complexity

Small organizations

Small organizations have simple structures, short lines of communication, and can be managed by one person who has a good

> Case 9.
>
> Up to a certain size, the comprehensiveness of an all-purpose authority unquestionably makes for better co-ordination between the services than is possible where they are divided between two levels oi local government. But beyond the point at which the size of individual departments begins to make communication between their staffs difficult, comprehensiveness itself becomes the enemy of co-ordination.
>
> D. Senior. *Memorandum of Dissent, Royal Commission on Local Government in England, 1966–1969.*

knowledge of what everyone is doing. Small organizations should have fast response times.

A small organization having fewer than, say, ten employees, may well be managed successfully by its leader giving instructions directly to each person. But when the number reaches perhaps fifteen or twenty, an assistant will be required to share at least some of the managerial tasks.

As the numbers increase further, the organization becomes divided into manageable units, each headed by someone in charge. This subdivision produces different levels, and the whole organization takes on a pyramidical form.

Subdivision also has an effect on the work being done in each level — the lower ones undertaking particular work (perhaps manual), middle levels carrying out planning, design, supervision, etc., and higher levels concentrating on general management, including policy formulation and finance.

Large organizations

Large organizations are likely to have greater resources available and be able to commit them to things like marketing, long-term research, and employee training.

The terms 'small' and 'large' apply to different organizations depending on the kind of activities which are undertaken. For example, an organization employing 400 people may be a small contractor or local authority, but a large firm of consulting engineers.

As organizations become larger, the person in charge no longer knows everyone nor controls everything personally, and the various tasks of management are carried out by subordinates under delegated powers. Specialist sections become necessary for activities such as work study, personnel management and finance.

In large organizations, employees tend to find it difficult to relate their activities to those of other parts of the organization. Increasing size makes internal communication more difficult. Inevitably therefore, large organizations are more formalized in the way in which they are structured than are smaller ones.

Examples of organization structures

Examples of organization structures to be found in civil engineering are given in Figs 3—6, but these are not to be taken as standards or models for everyone to follow.

Contractor

Figure 3 shows a structure for a medium-sized contracting firm engaged in civil engineering and building in both the UK and abroad. The structure includes a non-executive director (who holds a part-time appointment) to bring in an outside view of the company and the industry.

While the organization of every contractor's firm will be different in detail, they all need to have someone in charge, and this is the Managing Director. Sometimes he combines this role with that of Chairman, representing the owners, be they shareholders or a group of companies.

To a greater or lesser extent, depending on the size and specialization of the company, the following functions have to be 'directed' by the Board of Directors: corporate planning, marketing, estimating, constructing, measurement, plant, specialist activities, buying, design, costing, accounting, corporate finance, office management, wages and salaries.

Only in the largest companies will each function require its own section or department. But even the smallest construction firm is usually a three-man team dealing respectively with construction, accounting, and commercial and technical matters.

Figure 4, which is reproduced from *Civil Engineering Procedure* (ICE, 1986), shows organization structures for a contractor's site agent and for a resident engineer.

Consultant engineer

Figure 5 is an organization chart for a medium to large provincial consulting engineer's firm.

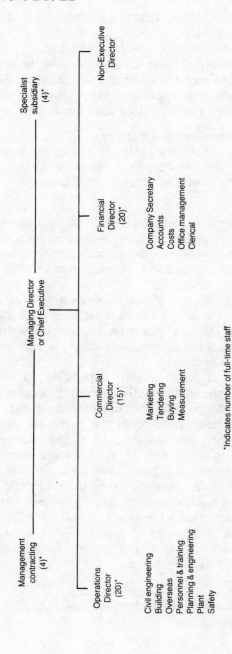

Fig. 3. Organisation chart for a medium-sized construction company's head office

It would be staffed mainly by chartered civil engineers, graduates and technicians, and although each division would be autonomous, interchange of staff would be encouraged so as to widen experience. The permanent staff would be supplemented as necessary by staff on contract to deal with fluctuating workloads.

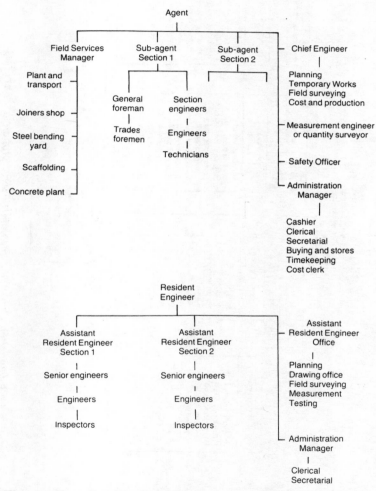

Fig. 4. Organization charts for a contractor's agent's staff and a resident engineer's staff

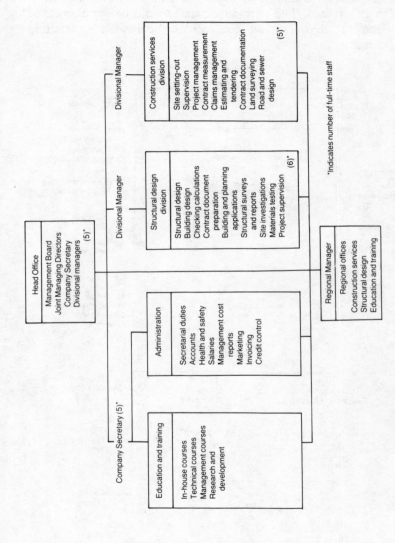

Fig. 5. Organization chart for a medium to large provincial consulting engineer

It would have its own micro-computer facilities, and such a firm would be able to offer a range of professional building and civil engineering services to all sectors of the industry.

Local authority

Figure 6 is an organization chart for a county surveyor's department in a large local authority, which is often the promoter or employer for construction projects. Typically there is a deputy to the Chief Officer, who would normally have specific responsibility for certain aspects of the department's duties.

There are three main divisions: design, administration, and maintenance and direct works.

Not shown on the chart are important links with other departments such as personnel, finance and legal staff, as well as with councillors and the public.

Principles of structuring

Organizations may be designed in various ways and Wearne has identified four systems for this purpose:

- *Phases of work* — in this system, each group of people takes responsibility for one phase of the work. For example, one group would undertake the feasibility study, another the design, a third, the detailing needed for giving instructions to the contractor, and so on.

- *Levels of experience* — this involves authority for making decisions being handed down a hierarchy of levels as work proceeds.

- *Projects* — by this method a group of people is formed to undertake a complete project.

- *Specialisms* — this involves forming groups of specialists to undertake certain aspects of work on a permanent basis.

Project basis

Grouping by project is a logical and productive means of organizing work. It brings into one place the expertise needed to undertake the whole of the work, thus reducing problems of communication and flow of information. People in a project team feel motivated by its objectives.

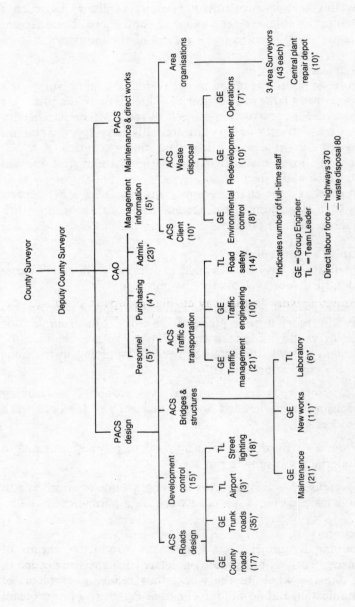

Fig. 6. *Organization chart for a large local authority surveyor's department*

Groups may be formed specifically for each new project, or a basic unit may remain in existence, supplemented as required by the size of a project or the need for different expertise or talents. Group leaders require not only specialist skills but also those of general management.

Specialist or subject basis

In this system people are grouped according to their expertise, e. g. in structures, highways, drainage, or soil investigation and analysis.

The advantage of grouping people in this way is that they can develop their own expertise to a high level. Knowledge can be accumulated and passed on to new members of the group.

A disadvantage is that such groups can evolve their own objectives distinct from thos of the organization, and integrating their work into projects involving several other groups is less easy.

Phase basis

Organizing work in this way enables groups of people to concentrate on a phase of decision-making with an immediate objective, as a project evolves. Thus one group may deal with feasibility studies, others with different stages of design, and yet another with implementation.

Controlling all aspects of work organized on this basis, as one phase succeeds another, may result in delay and lack of co-ordination.

Levels basis

As most organization structures tend to be hierarchical in form, it is natural that the decisions to be made and the work to be done at different levels will vary from simpler, more routine matters at lower levels, to more difficult and complex ones at higher levels. Thus, to some extent, all work will be allocated on the basis of levels in the organization structure.

Examples of organization structures in civil engineering

Wearne considers alternative forms of organization for undertaking work. The following three examples (in abbreviated form) are for a typical promoter, a consultant, and a contractor.

Promoter

This promoter is a public authority which has a statutory obligation to provide and maintain a municipal service. It is an example of organizing work mainly in phases, with some by levels and some by specialisms, and illustrates an organization which is suitable where projects require little innovation of either design or decision-making.

The project consists of civil engineering and building work and the installation of mechanical and electrical equipment. The operation of the completed installations will be almost entirely automatic. Innovations by the suppliers of equipment are utilized but the main decisions in the design are in adapting civil engineering experience to the site conditions.

Decisions begin from the promoter's forecasts of changes in demand. The performance and timing of the project are outlined and assessments made of when to close old installations which have become uneconomic.

On receiving instructions to proceed, there follows the main design of the project, which is not in sufficient detail to invite tenders. At the same time, specifications are prepared in order to invite tenders from manufacturers for the supply of mechanical and electrical equipment.

Before construction can begin, details have to be submitted to the departments and authorities for town planning and building regulation approvals.

The flow of design information and the co-ordinating role of the civil engineering section is shown in Fig. 7, reproduced from *Principles of engineering organization* (Wearne, 1973). In this diagram, as well as in Figs 8 and 9, the chain-dotted lines show the information flow in the first phase of design; broken lines, the flow in the second phase; and full lines, that in the main phase. The starting point of each is numbered.

After the construction is complete, drawings have to be revised to show all details as built, and instruction manuals prepared for both maintenance and operation of plant and equipment.

The parallel involvement of several sections requires planning of the flow of information, and linking comments with decisions. Achieving this is the responsibility of the section with the most work in the design of the project.

The main phase of the design is divided among specialist sections which work in parallel, and with outside consultants if necessary, where a specialism is lacking within the department.

A means of reviewing the effectiveness of the system is provided by regular meetings of the Chief Engineer and group leaders.

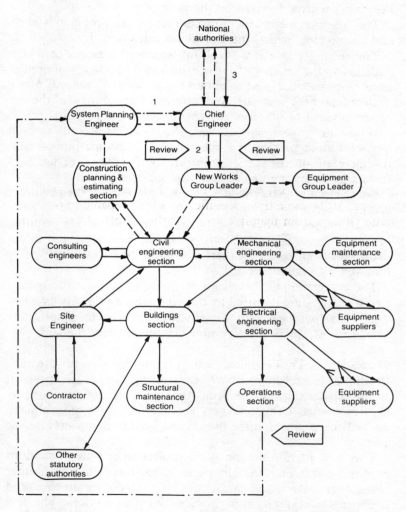

Fig. 7. The flow of design information in a promoter's firm

Consulting engineer

A firm of consulting engineers specializes in one branch of civil engineering. It undertakes feasibility studies, design, preparation of contract documents, site supervision and contract administration for promoters of projects, as required.

Design for a project divides naturally into four phases, each beginning with a decision by the promoter.

The first phase consists of discussion on the promoter's ideas and the services which the consultant can offer.

The second phase is a feasibility study, the extent of which depends on the urgency of the project and the accuracy of predictions of cost — required to obtain the finance to proceed. Many projects end here or are delayed, awaiting the purchase of the site or the consent of others who are affected.

When the project can proceed, the next phase tends to be a hurried one of making the main design decisions, but concentrating on producing sufficient information with typical details to form a basis for inviting tenders for construction.

Completion of the detail design takes place when the promoter accepts a tender.

In all phases of work for a project, the consultant's planning of design depends on the decisions of the promoter, and in the main phases of detailing there are likely to be cycles of delay followed by urgent working.

The completion of the design may have to be particularly hurried if changes are required by the promoter, and in any case this final phase overlaps with construction in order to adapt detail design to information on ground conditions from the site.

Partnership. The consultant's firm is founded on a private partnership of several engineers. The managing partner is a specialist in the typical work of the firm. Another specializes in construction problems. Other partners act individually as consultants, and all the partners share the task of getting new work for the firm.

The flow of information on a project begins in discussions between a promoter and any partner, but the managing partner takes over to advise a promoter on proceeding further and negotiates the agreement to provide the firm's services. The flow of design information is shown in Fig. 8.

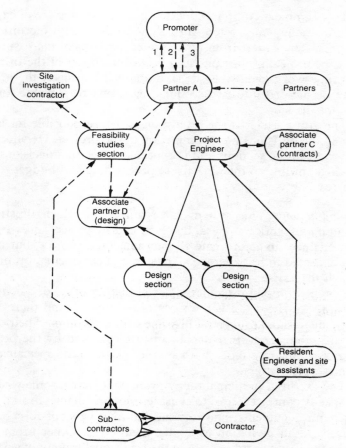

Fig. 8. The flow of design information in a consulting engineer's firm

The feasibility studies for all projects are carried out by one section of engineers. They specify the information needed from the promoter and site investigations, analyse alternative proposals, and prepare a report on their conclusions, working with appropriate partners.

When the promoter instructs the firm to proceed with the next phase of design and drafting specifications for construction, the consultants assemble a project group of engineers, draughtsmen and assistants to carry out this much greater volume of work.

To assemble a group rapidly, temporary use is made of experienced draughtsmen hired through an agency, with recruitment and transfer of staff from other projects gradually increasing so that the project group consists solely of members of the firm by the time work begins on-site. On a sizeable project, the group may be divided into two sections, each one responsible for an area of the site.

The group leader is the project engineer, responsible for linking all the decisions until completion of the project. He allocates the work to the sections and is expected to see that decisions are made according to the objectives, programme and budget set in the feasibility studies.

Co-ordination. This system of organization, which illustrates grouping of work by phase, by level, and by project, uses a project engineer to co-ordinate all decisions on a project, but he is not expected to be expert in all design and contractual problems which may arise.

His most important concern is the linking of all decisions on design, but the detail is the work of sections — and their decisions on design are reviewed by one of the partners. Their draft specifications and other documents are reviewed by the project engineer, in consultation with another partner who specializes in contractual matters.

The extent of participation by promoters varies, some wishing to specify contract terms, to consider the consultant's draft drawings and documents, and to issue the final versions to contractors, with an invitation to tender themselves. Other promoters leave the firm to act for them in the traditional way, only wishing to see the tenders after assessment, with a recommendation from the consultant.

After the contractor enters into possession of the site, the project engineer and his group are responsible for the completion of the design and for the administration of the contract. The resident engineer on-site becomes the link with the contractor, and he passes back to the group for approval, the drawings and other information supplied by the contractor showing the methods he proposes to use in the construction of the work.

This example illustrates grouping of design by phase, by level and by project, all within one branch of engineering.

Contractor

As a process plant contractor, this firm undertakes turnkey contracts to design and supply complete process projects or sections of them, but subcontracts to specialist firms the detailed design, manufacture and construction of all equipment and structures.

The origins of projects vary. The firm invests in some research into processes. The results, together with market predictions, are used to make feasibility studies to present to manufacturing firms who may be attracted by these proposals and so promote a project. An invitation to tender and negotiation of an order can follow.

But the firm tenders much more often for projects based upon processes already chosen by the promoters, or requiring the firm to negotiate to use a process developed by others. In these tenders, the firm is usually competing with other contractors, but once working for the promoter, may subsequently get the opportunity to negotiate to undertake further work.

Speed is usually important to the promoter and so development during design is not favoured.

The contractor needs to keep himself informed of innovations in equipment which are of potential use, so as to take advantage of them in offering tenders to promoters. Such innovations have to be proved by subcontractors before their products are used.

Tendering. The start of a project is usually the preparation of a tender, but it can also be in the prior study of a novel proposal, to demonstrate its economic value to potential promoters. In either case, the success of a tender in leading to an order is uncertain, and the firm limits its expenditure in these early phases of a project to the minimum necessary to estimate the probable cost of the project, and to satisfy the promoter that the performance proposed is likely to be achieved.

Decisions on the process, control systems and layout of the principal sections of the project are essential in tendering, but the design of structures and of chemical, electrical and mechanical equipment is considered in this phase only in so far as is needed to investigate novel problems, or to provide detail required by a promoter.

The projects group — composed of sections of engineers, com-

55

mercial and planning experts — specializes in the preparation of tenders. In consultation with all departments of the firm, they prepare programmes for carrying out the design, etc. of the project which will follow if an order is received. When available, the person who is to be project manager may be attached to the group during this phase, in order to take part in all the planning.

In this phase of decisions, specialists concerned with the process, performance, safety, and layout take part in the design decisions for tendering, but the main group of sections dealing with the design of equipment and structures may be consulted only.

The relationship between this main group and the other groups reverses when an order is received to build the project. The project now moves into the main phase of selection of equipment and design of linking services and structures. These tasks are divided among the several large design sections which mostly make parallel decisions.

The flow of information is shown in Fig. 9.

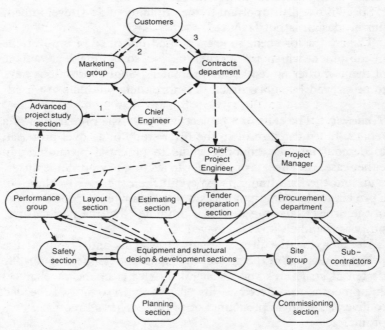

Fig. 9. The flow of design information in a contractor's firm

Specialists. In the sequence of decisions, the specialists in project sections work increasingly in parallel and are concerned with detail from other firms which must be brought together by them and must meet the project objectives of performance, time and cost already decided in the tender and negotiations with the promoter.

In this system, many people are grouped in sections specializing by subject. These are formed into three groups, two of which consist of sections primarily concerned with performance and other decisions for tenders while the third, much larger group, consists of those involved in the major phase of detail after the firm has an order from the promoter.

Thus, there is grouping by phase of a project, but sections in the larger group are also expected to make advance proposals for novel projects and they may be consulted during tendering for any project.

The Company Director

The responsibilities of a director are to his company, its shareholders, its employees, its customers or employers, and its creditors, as well as to fellow directors.

This is a definitive list and applies whatever the size or status of the company. Status extends from a public limited liability company with shares quoted on the Stock Exchange at one end of the scale, to a small private company at the other.

These responsibilities, which are to be found in the Companies Act 1985, sometimes conflict, and it is when an attempt is made to list them in an order of priority that problems can occur. Only one overrides all the others: a director's duty to his company is paramount.

A company director may be part-time and non-executive and can be an employee too. In either case he may be a shareholder, but the role of director is completely separate from any other function.

The shareholders are, of course, the owners of the company, and it is their investment for which the directors are responsible. The shareholders elect the board of directors by way of their votes at the general meetings of the company. The board of directors then elects from within its ranks a chairman and a managing director, sometimes named the Chief Executive.

The employees look to the company, and hence to its directors, to provide their means of livelihood not only at the present but in the future. To have a reputation as a 'good company to work for' should be a prized asset of any company. To achieve and maintain such a reputation is one of the many responsibilities of the board. They will do this by satisfying themselves of the quality and training of their managers and ensuring that fair employment policies are carried out. Incentives are often used to recruit and retain senior managers, such as commissions on profit and options on shares in the company, at some future time, at a favourable price.

Collective responsibility. To return to the axiom that the company is the key to all the Director's responsibilities, one of the complications that arises is that the Director can also be a shareholder, and often is a manager and employee of the same company.

At a board meeting, policy decisions must be made by the directors, who will have shed any other role they play in the company. They should not forget that they are voted onto the board by the shareholders to see that the company's business is operated at this maximum efficiency and yields an adequate return for the shareholders. The decisions they take will be their collective responsibility.

However, directors of a subsidiary company, such as for example, the well-known Balfour Beatty Ltd, are responsible to the board of the parent company, in this case BICC plc, who in turn are responsible to their shareholders.

The Director, who is also a manager, will, after the board meeting, have to ensure that his staff understand the policy decisions of the board and see that they are put into operation. This applies whether he agrees with them or not, and whether they are in his own interest or not, as an employee and, possibly, a shareholder.

However, what is good for the company is often good for the employees, and invariably good for the shareholders, so there is not usually any conflict. This however, often arises in a takeover situation; the shareholders will wish to sell to the highest bidder, but the new owners may make some senior executives and other employees redundant.

So far, the Director's duty to the company's customers has not been enlarged upon. The customers in the construction industry are usually called the Employer under the standard forms of contract, although sometimes called the Authority or Purchaser. Whatever they are called, they provide work, and no company, whatever its size, can afford to have poor relations with its customers.

Finally, there are the company's creditors, for whom there is a legal requirement for a company to reserve capital for their protection, i. e. so that they can be sure of being paid. But, in addition to this legal requirement, there is an overriding reason for maintaining fair credit terms: this is to ensure that supplies of goods and services are always available, with adequate time before the company has to pay for them.

Construction projects

There is a need for better control of work and progress of construction projects.

In *Control of projects during construction* (Ninos & Wearne, 1986) it is shown that the problems in practice lie in achieving control of a project as a whole when decisions on its objectives, financing, planning, design and construction are divided between departments in the promoter's organization, consulting engineers, architects, contractors and subcontractors.

The whole process for a major scheme can be complex, as is shown in Fig. 10, and the consequences tend to be that:

- problems are solved on the basis of each party's own objectives

- data tend to be held in a form which is not consistent with the needs of other users

- problems involving more than one party tend to be raised too late to avoid delays and extra costs

- these problems become the subject of disputes

- practice both within and between organizations becomes formalized, so that individuals believe that they cannot make improvements

- repeated failures lead to the imposition of procedures which themselves add to costs and make for delay.

59

Thus, the promoter should appoint one person to be in control of a project and to have command of the relevant technical and

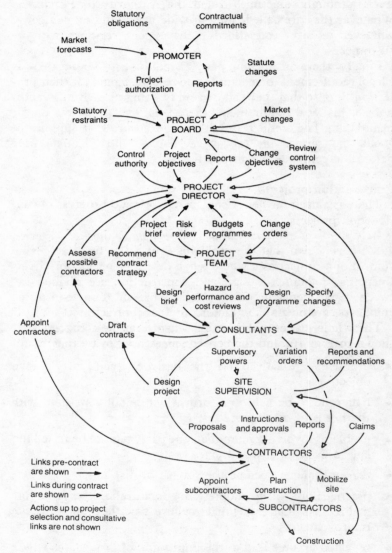

Fig. 10.   Responsibilities for construction

managerial information, and the time, authority and incentive to use it.

For major schemes, a project board may be required, and for public sector schemes this may be a steering committee of departmental heads.

Case 10.

Decisions are most successful if they are controlled by one person who is responsible for the results, and who has the authority, the time and the information to investigate the need for and consequences of changes proposed to design, programmes or budget.

Ninos & Wearne, *Control of projects during construction.*

Organizations in action

A useful summary of the most important factors affecting organizations in action is given by Wearne.

- No one system is right for all organizations or for all parts of an organization. The choice of structure and the extent of formal definition of jobs should be contingent upon the scale, culture, complexity, novelty and urgency of the primary task, i. e. the essential productive work.

- The tendency of organizations is to become mechanistic. This can be suitable for staff functions, but the line (*see* Chapter 4) employed on the primary task need more organic management. This is particulaly relevant to engineering organizations. Only predictable tasks can be structured. The tasks of engineers are, in general, increasingly unpredictable.

- Needs change during a project and during each major stage of work. Begin with organic relationships. Growth in numbers may require more structure, but revert to the organic to be flexible in problem solving for project completion.

- The basis of choice of the system of organization should be analysis of the relationships needed for the primary task, particularly the flow of information and decisions that start from external relationships with clients, bankers, government, etc.

- Remember:

o under old factory traditions, bureaucratic caution and staff influence, the tendency has been to over-organize

o autonomy supported by counselling can be most motivating, and plans and systems imposed on people are usually demotivating

o only line people (*see* Chapter 4) employed on the primary task earn the money to pay for everyone, so minimize the numbers in managerial and staff roles.

Further reading

Burns, T. & Stalker, G. M. *The management of innovation.* Tavistock, London, 1968.

Civil Engineering Procedure (4th edn). Institution of Civil Engineers, London, 1986.

Dressell, G. *Organization of a construction company.* Maclaren, London, 1968.

Handy, C. *Understanding organizations.* Penguin, London, 1987.

March, G. J. & Simon, H. A. *Organizations.* Wiley, London, 1958.

Ninos, G E. & Wearne, S. H. Control of projects during construction. *Proc. Instn Civ. Engrs*, Part 1, **80**, 931−943, Aug. 1986.

Pugh, D. S. *Organization theory.* Penguin, London, 1977.

Senior, D. *Memorandum of dissent; Royal Commission on Local Government in England, 1966−1969.* Volume 2, Cmnd, 4040 I. HMSO, London, 1969.

Sofer, C. *Organizations in theory and practice.* Heinemann, London, 1972.

Wearne, S. H. *Principles of enginering organization.* Edward Arnold, London, 1973.

Weber, M. *The theory of social and economic organization.* Oxford University Press, Oxford, 1947.

4 Management and supervision

Principles

For many years, there has been a quest for general principles of management which would assist in making organizations function better. The best known principles were prepared, after many years experience in large organizations, by Henri Fayol, a mining engineer. He maintained that principles of management should not be regarded as in any way rigid, but should be flexible and adaptable to every need. They can be seen in use, or their absence noted, in organizations day-by-day.

Briefly, Fayol's principles are as follows.

- *Division of work* — a worker who specializes in a particular activity is likely to become expert, and therefore will produce more with the same effort.

- *Authority and responsibility* — there are two kinds of authority: that deriving from position (job title and level in the hierarchy); and that which comes from the abilities and personality of the individual. The former is not sufficient on its own for successful management; thus personal authority is the indispensable corollary of official authority. Responsibility is an essential counterpart of authority; thus wherever authority is exercised, responsibility rests.

- *Discipline* — this is necessary for the smooth running of organizations, and consists of obedience, application, energy and behaviour. People are obedient and loyal if they are ably led. There are three requirements for good discipline:

o good superiors at all levels
o agreements should be clear and fair
o penalties should be applied judiciously.

- *Unity of command* — a person should receive instructions from only one superior, otherwise authority is undermined, discipline placed in jeopardy, order is disturbed and stability is threatened.

- *Unity of direction* — this refers to there being one head and one plan for a group of activities having the same objective; it is necessary to ensure unity of action, the co-ordination of strength, and the focusing of effort.

- *Subordination of the individual interest to the general interest* — ignorance, ambition, selfishness, laziness, weakness and all human passions tend to cause the general interest to be lost sight of in preference to the individual interest. But for the success of the organization, the interest of one person or group of persons should not prevail over that of the organization.

- *Remuneration of personnel* — this should be fair and satisfactory to both the person and the organization. Effort should be rewarded to encourage keenness.

- *Centralization* — the amount of centralization or decentralization is a matter of degree, and varies from time to time in any organization. The former will tend to reduce the role of the individual, and the latter to enhance it.

- *Scalar chain* — this is the invisible chain which links those in authority, from the highest to the lowest levels. In theory, all communications to different levels travel up and down the chain. In practice, communication is usually made directly from one level to another level, without going to the top of the chain each time. This is satisfactory provided that decisions made and action taken are acceptable to superiors.

- *Order* — material order consists of having well selected places for the storage of materials and ensuring that orderliness prevails. Social order consists of deciding which posts are needed and filling them with appropriate people.

- *Equity* — people desire equity and equality of treatment.

- *Stability of tenure* — people need time to learn new work and, having done so, should remain in employment to give worthwhile service.

- *Initiative* — this can be a great source of strength to all organizations and should be encouraged and developed in everyone.

- *Esprit de corps* — harmony is a great strength in an organization and efforts should be made to establish it. To that end, personnel should not be split up, and over-use of written communication avoided when oral communication would suffice.

Minimum division of work

Division of work into discrete parcels which are allocated to separate groups will tend to inhibit co-operation between people and thus hinder the achievement of the organization's objectives.

If division of work is combined with geographical separation, the flow of information is retarded, and the exchange of ideas and discussion of potential solutions to novel or complex problems is rendered more difficult.

Span of control

The span of control, or span of management, as it is sometimes known, refers to the number of people who are directly responsible to one person.

The number of people who can be supervised effectively depends on a number of factors, including the capabilities of those involved, the complexity, familiarity and predictability of the task and the extent of delegation.

Someone who has been promoted through an organization in which he has worked for a long time will be very familiar with the organization and its people and systems, and therefore may have a wider span of control than a recently appointed person at a similar level.

In practice, span of control varies widely, although a figure of around six persons is typical. The wider the span of control, the fewer organization levels there will be, and the greater the

Case 11.

With the reorganization of local government in 1974, the departmental heads of the new district councils were given a unique opportunity to design organization structures.

One director of technical services who was appointed head of a multi-disciplinary department of engineers, architects and town planners in an authority for some 100 000 people, and who was well-versed in the theory and practice of management, carefully considered his span of control.

Bearing in mind his need to attend several committees regularly, to be a member of the Chief Officer's management team, and to be available for consultation by councillors and members of the public, he wished to limit his span of control to three.

Consequently, he appointed two deputy directors, one for engineers and architects, and the other for town planning, as well as a principal administrative assistant, who would be responsible for the annual budget, procurement and wages.

One of the deputy directors had four section heads:

- *Highways* – county council agency for maintenance and minor improvement
- *Drainage* – water authority agency for sewerage, including pumping stations, and for land drainage
- *Architecture* – design and construction of public buildings and houses; estate layouts
- *Building control.*

The other deputy director was in charge of three sections:

- *Local plans*
- *Development control*
- *Recreation* – swimming pools, parks and open spaces, sports centres.

It was not long before the section heads at third tier level complained of general dissatisfaction, in that the chain of command was too long, and the Director too inaccessible.

Equally, the Director found himself out of touch with the thinking of section heads and their day-to-day problems, and felt a general lack of knowledge of what was happening.

When a vacancy occurred for one of the deputy posts, it was kept unfilled and the section heads in that division reported directly to the Director, thus doubling his span of control.

It took some time for the section heads to absorb their new responsibilities at the higher level, and for a while it took up time which the Director could ill spare, but everyone was intent on making it succeed, which it did within a year.

> *Case 11 — continued*
>
> Consequently, when a vacancy occurred in the other deputy post, that too remained unfilled and the section heads undertook the duties of the post between them.
>
> The result was more delegation to the section heads, higher salaries and more job satisfaction. The span of control of the director increased from three to eight, but he was better informed and ran a happier department. The council got a better service at lower cost.

amount of delegation to subordinates, and thus of job satisfaction for them.

It must be remembered, however, that not all employees seek more delegation or more responsibility.

An illustration of how span of control can change is given in Case 11.

Line and staff

A 'line' role is responsibility for the primary task.

A 'staff' role is an advisory one, but within a staff section there may be line relationships.

Staff employees are usually to be found in separate departments which give support to the primary task, for example by dealing with things like research, personnel matters, and public relations.

From time to time, those who are usually line managers may find themselves in a staff relationship with other managers. In any particular situation it is important for each person to understand whether he is in a staff relationship (giving advice), or a line relationship (giving instructions).

Bypassing subordinates

In a formal type of organization, one of the principles which should apply is unity of command, by which each person receives instructions from one person only. Typically, organization charts show each post holder responsible to one at a higher level, yet in practice, day-by-day, employees may receive instructions from more than one person. This is not always satisfactory and can lead to confusion, uncertainty and resentment.

This problem is one of many examined by Wilfred Brown (Chairman and Management Director of the Glacier Metal Company) and Elliott Jaques, over a long period of research into every aspect of organization in an industrial enterprise.

> **Case 12.**
>
> A construction company is interested in a project overseas and the Managing Director agrees with his operations and commercial directors that it would be worth incurring the considerable expense of tendering.
>
> The tender team will be headed by the Overseas Manager, and will include the person who will be the Agent or Project Manager if the contract is won, together with the Senior Estimator allocated to the tender preparation.
>
> In addition, the Plant Manager, or his representative, will be in the team, as it is expected that maintaining the plant will be difficult in the arid territory in which the project is situated.
>
> During tender preparation, there is a staff relationship between the Overseas Manager, the Estimator, and the Plant Manager; each one being responsible for his own specialist input to the tender.
>
> During the visit abroad, the tender team forms a matrix type of organization, with the Overseas Manager taking responsibility for the whole team's actions while out of the UK.

Brown defined the immediate command of a manager as that group of employees which is directly accountable to him (in effect, his span of control) and his extended command, as all the employees under his control.

Sometimes a manager will need to communicate directly with an employee in his extended command, instead of through one of his immediate subordinates. Having done so, he will have bypassed the person who is the employee's superior, and thereby removed responsibility from him for that particular instruction.

The manager should therefore inform his immediate subordinate of his action as soon as possible, and ensure that the employee is returned to the command of his superior when the task is complete. In this way, some of the difficulties that would otherwise arise from bypassing subordinates can be avoided.

Collateral relationships

Collateral relationships are those which exist between people of the same rank, whose activities impinge on each other. The effectiveness with which they operate concerns both of them, though each one is responsible only for his own section or department.

Difficulties that arise should be settled between them, but when this is not possible, the disagreement should be referred to their superior. It is his task not so much to arbitrate, as to try, through joint discussion, to guide his subordinates to a solution which will emerge when all the relevant factors are disclosed.

While there is nothing inherently wrong with constructive conflict or fruitful competition, there can be danger for the organization in arguments about territory and roles which stem from personal ambition. They are less likely to occur where there is a healthy environment and good leadership.

Matrix principle

Under this system, people become responsible to a second superior, say a team leader, for a particular purpose, and are thus responsible to more than one person (their usual superior and the team leader) but to each for a different task.

It is often necessary in the construction industry to bring together a team of specialists from different sections or departments to design a scheme, and one may consider this to form a matrix type of organization.

Members remain with the team until their contribution is complete, some working on it part-time, and on completion all will return to their normal work. While working with the team and taking instructions from its leader on the scheme, each member will remain under the control of his usual superior for such things as pay and promotion.

The matrix type of organization is flexible and adaptive and, through it, members can gain first-hand experience in a multi-disciplinary environment.

Information and communication

Activity in organizations is not possible without an exchange of information, which must be initiated, transmitted and received before action occurs.

As well as internal communication, organizations need to communicate with clients and customers and sometimes with the public too.

Communication is a two-way process, being both given and received, and it may be oral or written. The information being passed should be accurately and concisely expressed and should be relevant to the recipient's needs.

Information passes up and down an organization from superior to subordinate, and vice versa, 'through channels', as well as passing across it between peers.

Informal organization

Within every formal organization there exists an informal one. Indeed, in many ways the informal organization gets things done. Each incumbent of a post has his or her own abilities, and some will expand the duties of their posts by those abilities. Overlapping responsibilities, or gaps in duties which are not clear on organization charts or job descriptions, will be clarified or filled by zealous individuals. As employees come and go and relationships change, so the informal organization changes.

One manifestation of the informal organization is the grapevine. Information of interest to employees is passed along the grapevine, and it can be used by management for this purpose, it being quick though not always accurate. Often, more attention is paid by employees to the grapevine than to officially published information.

Delegation

As no one person can make all the decisions in an organization of any size, authority for making decisions must be delegated. This process involves investing a subordinate with such authority.

Case 13.

The reaction of individuals to events may be unexpected.

In one large organization, the head of a division failed in his bid to be promoted to the vacant post of head of the department. He was so disappointed and upset that he vowed to do the minimum of work necessary to keep his division ticking over, and thereafter devoted his energies to running the local amateur dramatic society.

The head of a small department in the same organization was told some time later that his department was to be amalgamated with a larger one and, as a result of this reorganization, he would drop from first to third tier level in the combined department.

He too initially decided to do the minimum amount of work, but within a week or two he realized that, for his own satisfaction, if for no other reason, he would have to continue to give of his time.

The superior will wish to make it clear that, with the authority, there also goes responsibility, which in practice means that the subordinate is accountable to his superior for his actions.

Both parties need to be quite clear about the extent of the delegation. The process of delegation has been defined as:

- the determination of expected results
- the assignment of tasks
- the delegation of authority for accomplishing these tasks
- the exaction of responsibility for their accomplishment.

The extent of delegation may be specific and set out in writing, or it may be expressed in general terms. More often than not it is simply assumed by both superior and subordinate, which works well enough until things go wrong.

Human nature

The way in which managers manage and supervisors supervise depends to an extent on their perception of the people working under their control.

Douglas McGregor describes the assumptions about human nature and behaviour which lie behind managerial decisions and action in his Theory X, the traditional view of direction and control.

- The average human being has an inherent dislike of work and will avoid it if he can.

- Because of this characteristic of dislike of work, most people must be coerced, controlled, directed, or threatened with punishment to get them to put forth adequate effort toward the achievement of organizational objectives.

- The average human being prefers to be directed, wishes to avoid responsibility, has relatively little ambition, and wants security above all.

McGregor went on to formulate a number of generalizations that he believed provided a modest beginning for a new theory about the management of human resources, which he called Theory Y.

- The expenditure of physical and mental effort in work is as natural as play or rest.

71

- External control and the threat of punishment are not the only means for bringing about effort directed toward organizational objectives — people will exercise self-direction and self-control in the service of objectives to which they are committed.

- Commitment to objectives is a function of the rewards associated with their achievement.

- The average human being learns, under proper conditions, not only to accept, but to seek responsibility.

- The capacity to exercise a relatively high degree of imagination, ingenuity, and creativity in the solution of organizational problems, is widely, not narrowly, distributed in the population.

- Under the conditions of modern industrial life, the intellectual potentialities of the average human being are only partially utilized.

McGregor concluded that the assumptions of Theory Y illustrate that the limits of human collaboration in the organizational setting are not limits of human nature, but of management's ingenuity in discovering how to realize the potential represented by its human resources. If managers believe that the assumptions of Theory Y are valid, they should be taken into consideration in their day to day work.

Having studied 250 corporate managers in American organizations, M. Maccoby identified four types of character which he found in different combinations:

- *Craftsman* — an individual interested in making something. Self-contained and exacting, he can become unco-operative at times. He is a good master to apprentices but a poor member of a team. He leads by ordering subordinates to do what he decides upon.

- *Jungle fighter* — he needs power. He sees life as a battle for survival in which winners destroy losers. At his best he is a lion, protective towards his 'family' and ruthless with his competitors. However, his domineering attitude upsets independent subordinates and his obsession with defence can create unnecessary enemies.

- *Company man* — this kind of person is orientated to the institution; concerned with the human side of the company and committed to maintaining corporate integrity. He can turn into a senseless careerist, obsessed with organizational politics, but at his best he can sustain an atmosphere of discipline and service. He is perhaps too conservative to lead an innovative organization.

- *Gamesman* — he likes to take calculated risks and is fascinated by techniques, new methods and problems. He thrives on competition and can communicate his enthusiasm to his group. He is a team player who looks for glory rather than riches. At extremes, he can live in a fantasy world, finding games and competitions where none need exist.

All employees can see such characteristics in their colleagues, which may help them to understand other people's attitudes to problems and opportunities, as well as their own.

Leadership
We all exercise leadership from time to time; at work, during recreation, and in the home.

Early theory had it that leaders were born, not made, it being thought that the quality of leadership depended on inherited traits, or characteristics which were not widely distributed in the population.

Trait theory takes the view that successful leaders possess certain characteristics, and therefore, if these can be identified, they can either be developed in individuals, or those who possess them can be selected for leadership positions.

Characteristics such as sincerity, integrity, drive, ambition and initiative are considered essential in leaders, but others including decisiveness and ruthlessness have also been considered to be necessary in some circumstances.

However, it has been observed that not all good leaders possess all the characteristics deemed necessary, and that different situations demand different attributes. Nevertheless, much selection of people is still done on the basis of the assumptions implicit in trait theory.

Style theory is based on the belief that people respond better to some management styles than to others, it being assumed that

---

Case 14.

A procurement officer was asked to replace typewriters with word processors and he spent a lot of time carefully evaluating different machines which were demonstrated to him by various suppliers. Eventually he came up with a best buy and placed an order.

Unfortunately, the typists disliked the machines they were given and productivity fell and morale declined.

Eventually, suppliers were invited to give demonstrations to the users, who were able to try different machines. The users selected the type they preferred, which were ordered and supplied. After this, both productivity and morale improved.

---

people are more likely to follow a leader having a democratic style, and thus work more effectively, than one with an autocratic style.

But some people prefer to receive a greater degree of direction than others, and some situations (for example, emergencies) demand a style which leans more to the autocratic/dictatorial than to the democratic.

By and large, a supportive style is likely to produce a better response, more commitment and more involvement with the group.

Contingency theory, which is outlined below, takes note of the fact that there are many variables affecting leadership, and neither trait theory nor style theory satisfactorily encompasses those variables.

Management styles

Huneryager and Heckmann identified four main leadership styles:

- *Dictatorial* — work gets done through fear, by threats of penalties or punishment. This can achieve results, but there is doubt about their quantity and quality in the long term, and subordinates are usually dissatisfied.

- *Autocratic* — authority and decision-making are centred on the leader; participation in decision-making is not allowed, nor is deviation from instructions. This can achieve results, but they depend heavily on the leader's abilities.

- *Democratic* — authority and decision-making are decentralized; participation in decision-making is encouraged and leader and subordinates tend to work as a whole. This offers more promise than other types, but requires leaders of better quality.

- *Laissez-faire* — subordinates establish their own objectives and make all decisions; the result is usually disorganization or chaos, because individuals can proceed in different directions.

The style which individual managers prefer to adopt will depend on his or her own aptitudes and experience, the problems and opportunities being faced, and his perception of his subordinates.

Work on new patterns of management was carried out by Rensis Likert, who identified four systems of management pattern:

- System 1: exploitive authoritative
- System 2: benevolent authoritative
- System 3: consultative
- System 4: participative-group.

Analysis of a very detailed questionnaire completed by a group of managers, about their own organizations, revealed that most of them believed that their organizations corresponded to either System 2 or 3. When asked which System they would like their organizations to have, virtually everyone opted for System 4; the participative-group system.

It is claimed that the adoption of System 4 will be accompanied by long-range improvements in productivity, labour relations, costs and earnings. Likert concluded that a science-based management system such as System 4, is appreciably more complex than other systems, requiring greater learning and much greater skill to use it well, but yielding impressively better results.

Complexity

Vroom identified five leadership styles, broadly from autocratic to democratic, and defines seven questions that need to be answered for each decision that has to be addressed.

Case 15.

The Chief Technical Officer had a largish department and, in order to get to know his staff and manual workers, he kept Thursday afternoon free every week, when he would visit one section of 20 or more people in their offices or on-site.

He was able to get to know people and they him, and there were often new people to meet since most appointments were delegated to senior staff.

After he retired, he would pass the time of day with any of his former colleagues whom he chanced to meet in the street. On one occasion, two years after his successor took over, he stopped to talk to a gang spreading salt on town centre footways, who told him that they had never met, or even seen, his successor and would not know him if he walked past them.

- Is one decision likely to be better than another?
- Does the leader know enough to take it on his or her own?
- Is the problem clear and structured?
- Must the subordinates accept the decision?
- Would they accept *your* decision?
- Do subordinates share your goals for the organization?
- Are subordinates likely to conflict with each other?

While this is a neat and logical system, it appears complicated, and yet it only covers part of the work of a leader, thus demonstrating the complexity of the art of managing people.

Contingency theory

Fiedler's theory implies that the ability of a leader to exercise influence on a particular situation depends on the task, and the extent to which the leader's style, personality and approach, fit the group.

He identified three critical dimensions of the situations which affect the most effective leadership style: position-power; the task structure; leader—member relations.

In its simplest terms, his contingency theory states that the performance of a group will be contingent upon the appropriate matching of leadership styles and the extent to which the situation itself gives the leader influence over the members of the group.

Best-fit

Charles Handy described the best-fit approach as one in which it is recognized that there is no such thing as the 'right' style of leadership.

In any situation confronting a leader, there are four sets of influencing factors that he must take into consideration:

- the leader — his preferred style of operating and his personal characteristics
- the subordinates — their preferred style of leadership in the light of the circumstances
- the task — the job, its objectives and its technology.

These three factors and their fit will, in turn, all depend to some extent on:

- the environment — the organizational setting of the leader, his group and the importance of the task.

This approach to leadership suggests that there is no one style of leadership which is correct in all circumstances, and that leadership will be most effective when the requirements of the leader, the subordinates and the task fit together well.

Site management

Management of construction sites presents particular problems; those on large sites differing from those on smaller ones. It is usually assumed that management style is as important a determinant of the level of productivity on-site as it is in the office, and that productivity will increase with job satisfaction, better morale and higher levels of motivation. Thus, to increase these factors will improve productivity.

In *Leader orientation of construction site managers* (Bresnen, *et al.*, 1986) the authors discussed the effect of management style and other factors on site management, in the light of their research. They recognized that every job is different, not only in scale and type of project, but also in the nature and composition of the site team, the extent of subcontracting, relationships with client and designers, and the structure of the organization.

A manager's style may be broadly people-oriented or task-oriented, and the group found that on smaller, shorter, and mainly subcontract projects, variability in site managers' orien-

> *Case 16.*
> 'You get different styles of management. I think a fast job, when there's no room for messing about, you get a much more directive style of management. I think if you get time on the job, it's a much more consultative style of management, you are more willing to get blokes in to sort the thing out, be together, you've got time for a joint approach.'
> Bresnen *et al. Leader orientation of construction site managers.*

tation appeared to have little effect on productivity, whereas on larger, longer, and mainly direct labour sites, variability in their orientation appeared to be more critical.

Only on projects of longer duration was an orientation towards the development and maintenance of interpersonal relationships found to be significantly associated with the levels of performance which were achieved.

Their work suggests that leader orientation is important only in those situations in which members of the site team are largely directly employed, and consequently where relationships are more direct and continuous. Where subcontract labour forms the bulk of the workforce, it appears that variability in leader orientation has little effect.

The group considers that extensive use of subcontracting by main contractors may suggest a different role for site management, namely one having an emphasis rather more upon the planning organization, and co-ordination of work undertaken by contracted personnel, than upon the traditional man-management associated with work on-site, i. e. the directive or initiating aspects of leader behaviour.

Change

The management of change usually involves those comparatively small things which management wishes to introduce. Occasionally it involves setting up completely new departments or authorities, as in the case of the reorganization of local government in 1974, or the taking over of another firm.

There is often resistance to change because people prefer stability, do not wish their work pattern to be disturbed, and fear that they will lose power, income, or career prospects consequent upon any change.

Employees are more likely to accept technical change in the way in which work is done, than social change that affects status and relationships within the organization. Yet the health of an organization and its ability to compete in the market, or provide services effectively and efficiently, depends on employees being willing to adapt to changes in the environment in which it operates, and to internal changes necessary for it to remain viable.

Managers should encourage employees to welcome change as part of a continuing strategy for survival.

The President of one large American company set down criteria which can make change more acceptable to employees:

- when it is understood rather than when it is not
- when it does not threaten security rather than when it does
- when those affected have helped create it rather than when it has been externally imposed
- when it results from an application of previously established impersonal principles rather than when it is dictated by personal order
- when it follows a series of successful changes rather than when it follows a series of failures
- when it is inaugurated after prior change has been assimilated rather than when it is inaugurated during the confusion of other major change
- if it has been planned rather than if it is experimental
- if it is directed at people new on the job rather than at people old on the job
- if it is directed at people who share in the benefits of change rather than at those who do not
- if the organization has been trained to accept change.

Everyday management
Attitudes

The various chapters in this book deal with motivation, organization, structure, leadership, and all those other matters which are the concern not only of managers but of all employees, because from time to time almost everyone is involved in managing, i. e. getting things done through other people. How does all this sometimes complex knowledge help one in the day-to-day task of managing?

In the construction industry, it is not unusual for the young engineer at the start of his career to find himself directing the work of a gang of men who are undertaking a project on-site. At the same time he may have had to liaise with staff at the same level as himself to obtain transport and equipment, and finally would have received instructions from his superior about the project; its objectives, time scale and cost limits.

Thus, in this one project he will have adopted three basically different attitudes: receiving instructions from a superior; seeking and obtaining co-operation from peers; giving instruction to subordinates. And so it is with managers at all levels. There is always someone to be accountable to, equals to liaise with or persuade, and those to whom one gives instructions.

Giving instructions

The term 'giving instructions' conveys a quite inadequate idea of what can be a very complex matter. How, for example, can one reconcile the need to give instructions about a job, with the expectation that the subordinate will exercise responsibility in carrying it out?

The word 'subordinate' itself causes resentment because it implies that one person is 'under' another. People neither like being told what to do, nor being made to feel subordinate to another, yet in most organizational structures, some posts are below others.

In many instances it will not be necessary to give instructions at all; a particular situation, problem or aspect of a project can be discussed by superior and subordinate and a satisfactory solution will emerge if both have approached the task with open minds.

Face to face discussions are always likely to be more effective than instructions from afar, as the effectiveness of instructions is generally in inverse proportion to the distance they have travelled.

However, it will be necessary to give instructions from time to time, and this should be done with the courtesy expected between equals, always remembering that subordinates are not, and do not expect to be treated as, inferiors. They will often know more about a particular skill or craft than the superior. Neither is it a question of familiarity (by the use of Christian names, for instance); it is a matter of recognizing a person's potential to make a worthwhile contribution.

If it ever becomes necessary to exercise one's authority to get something done, then the manager has lost control of that particular situation.

Responsibility

One of the most frequent complaints of subordinates is that superiors do not delegate enough, and will not let go of tasks which subordinates feel quite capable of doing. The superior's desire to continue doing that which he did well and which earned him promotion, may be due in large measure to the fact that his new role has not been fully explained to him, or that he has not been trained for it, and that he still derives satisfaction from doing his previous work, at which he was adept.

Top management needs to ensure that people appointed or promoted to higher levels have had adequate training, but it should be remembered that individuals themselves have a role to play in preparing for the next step up the ladder — through self-development.

Among other things, promotion to a higher level will involve loosening control over detail, while continuing to show interest in the work of subordinates and encouraging them to exercise their skills and ingenuity to achieve an objective set in discussion with the superior.

But if one is to delegate, and most managers have to, staff must be competent and reliable. Thus staff selection, training and development must be well managed.

Courtesy

Managing people at higher levels in the organization differs somewhat from that at lower levels; there will be more concern with broader issues and perhaps with 'marketing' one's department or firm with colleagues and customers. Whatever one's role, the basic principles of consideration, tact, courtesy, and a recognition of the existence of other points of view, remain of foremost importance.

There is no ideal solution to the many different situations encountered in managing people, but basic knowledge is required in those who manage other people; how that knowledge is applied will depend on individuals and the situations themselves.

Supervision

The original, literal meaning of the verb 'supervise' is 'to oversee', but the dictionary definition nearest to the theme of this book is 'to watch over so as to maintain order'. The concept of order has already been mentioned earlier in this chapter, from having well selected places for the storage of materials, to fitting the right people into the right jobs.

In construction, the maintenance of order is as important as in any industry, not only in the field, but in the workshops and offices. It is, however, more difficult to achieve, and not only for the obvious reasons arising out of new projects producing new sets of circumstances.

It is more difficult because, in construction, one decision does not always produce one solution, but often leads to a situation where another decision is required. Thus the decision-making process is spread and is evident at all levels of supervision.

There is another aspect of construction that sets it apart from many other industries and that is the greater impact that excellence in one person can have on the entire project or business. This is particularly so on the construction site.

These are just some of the many reasons why supervisors must be of the highest quality and must be given the maximum of help.

Case 17.

The promoter of a pumping station decides that the capacity should be increased, and the designer makes the necessary changes to the permanent work, including deepening the pump chamber.

The contractor, who is well into the excavation, examines the temporary works and decides that the sheet piled cofferdam already constructed is sufficient to deal with the extra depth, provided that an additional frame of supporting walings and struts is introduced.

Site supervision will be involved in further decisions, not only in placing the frame in the cofferdam, but in dealing with the complications to the reinforcing steel and the shuttering of the concrete that the extra frame introduces.

With so many changes and so much progress in the management tools available, it is perhaps surprising that a theory propounded in 1955 has stood the test of time and is just as relevant today. It was proposed by P. F. Drucker, an American management consultant, and was called 'management by objectives'.

Improving the performance of supervision

As defined by Drucker, the principle of this technique is that the goal of each manager must be defined as the contribution he has to make to the success of the organization of which he is a part. He considered it important that the manager should develop and set these objectives himself, subject to approval from higher management.

This principle was developed and somewhat watered down by subsequent writers on management, with objectives being formulated by higher management with the involvement of subordinates such as supervisors.

The idea is that targets thus agreed are themselves an incentive and they form a yardstick against which performance can be measured.

The following table, from *Improving management performance* (Humble, 1965) defines the help that supervisors require.

The supervisor's needs

Needs	Methods available to help
'Tell me what you expect from me'	Setting objectives Key results analysis Performance standards Short-term action plans
'Give me an opportunity to perform'	Organization planning
'Let me know how I am getting on'	Control information Management performance review
'Give me guidance where I need it'	Management education and training
'Reward me according to my contribution'	Salary structure Succession planning

Setting objectives

The phrase 'tell me what you expect of me' is the supervisor asking his superior what are/is:

- the overall purpose of his job
- the key results to achieve this purpose
- the performance standards and control methods which relate to these key results
- the limitations of his authority
- the short-term priorities to integrate with the project or the department's needs and which, at the same time, satisfy the company's or authority's requirements.

Key areas are formally defined as those where excellence will have the most impact on the business or project, and those where poor performance would threaten it significantly. Performance standards are both quantitative and qualitative and relate to the job, not to the job holder.

Organization planning

Does the organization make proper use of human resources? These are some of the questions that need asking.

- Is there an effective division of the work to be done?
- Are responsibilities and objectives really understood?
- Is there a sound line of command?
- Is there provision for control, accountability and order?
- Is the organization working in a good spirit?

The answers to these questions will highlight weaknesses which must be corrected if they are not to present obstacles to high performance.

No supervisor can perform without the necessary resources. On a construction site these will include suitably skilled manpower, materials, plant, and clear instructions in the form of drawings, schedules and programmes.

Control information

This is usually concentrated in comparing progress with programme, and cost with budget. Quality must not be forgotten and is an important component of control information.

When setting up the cost system, the selection of operations to

---

Case 18.
Objectives for a section foreman — structures

The overall purpose of this appointment is to assist the Project Manager or Agent of the project to which you are assigned to achieve the contract completion date within budget and to the required quality.

On the Motorway Contract M28/5 you will supervise bridges B1,B2,B3 together with culverts to chainage 3000. This will be followed by allocation to other structures on this project or elsewhere.

The short-term plan is to complete sufficient of the above structures for the carriageway concrete train to be able to carry out its work over or under these structures by the dates specified in programme C104 or subsequent revisions.

The key to achieving these targets will be your team, especially the carpenters. A bonus scheme will operate, the formulation of which you will be involved in. You will also be responsible for the weekly returns of output.

In the longer term, the Company must satisfy the Employer and the Engineer that they are suitable contractors to carry out the building of future projects.

---

be measured is vital. Non-repetitive or one-off items, small quantities or less important work is omitted or collected together so that the operations that matter are not cluttered among unimportant data. The items that remain will then get the attention they deserve.

Performance review or appraisal

In an industry such as construction, with each project presenting different circumstances and problems, review is part of the routine of every manager, almost daily.

However, formal appraisals are part of management by objectives and are dealt with in Chapter 5. The problem in the construction industry is that, because of fluctuating workload, staff are often obliged to move from one employer to another, and this is particularly prevalent at first line supervisor level.

Management education and training

This is also dealt with in Chapter 5, where the importance of self-help and in-career courses are discussed. In the context of supervision, much of the development of the skills and knowledge required to improve performance will be acquired on the job, under the guidance of the best management in the right environment.

Salary structure and career planning

Salary structure involves setting the right level of pay for the job responsibilities, and also includes payment by results, such as bonus and commission.

Career and succession planning is difficult in the construction industry for the same reasons as with appraisals. However, it is important to have long-term, as well as short-term, plans for the training and development of supervisory talents to meet a predicted scale of requirements.

Supervising at the sharp end — the first line supervisor

Whether he is a section leader in a design office, a foreman on a construction site, or a chief clerk in the office, the first line supervisor is the key to bringing out peak performance from those working for him.

Yet, while management preaches that the supervisor's first duty is to human relations, there is a tendency to promote a supervisor for keeping good records and making life easier for his superior. The supervisor's problem is often that he has so many things to do, without knowing which are important. He therefore needs clear-cut objectives and the authority that goes with the responsibility for reaching those objectives.

Each individual should know how his work contributes to the whole; he should know the objectives he has to meet and how he is doing in relation to them, and he should be able to take a pride in his accomplishments.

Supervision of the workforce on the construction site

The two forms of managerial control exercised by construction companies for supervising the workforce consist at one extreme, of a strategy of trust, with the onus and responsibility for the speed and quality of the work placed upon the employee. This can only be done after recruiting and selecting workers who are thought to be reliable and trustworthy, and with whom supervisors can build up a permanent working relationship.

The other extreme is to subcontract the work so that the

responsibility for supervising other than quality and safety lies elsewhere.

Normally it is a combination of these two extremes, together with some intermediate stages, that are found on construction sites in the UK. Specialist operations are these days invariably subcontracted. The basic skills such as bricklaying, shuttering, and fixing reinforcement are either carried out by subcontractors (usually suppliers of labour only) or by directly employed operatives.

The combination depends on the state of the market, for if a contractor can see continuity of work in one area in one trade, he will employ his own men. It also depends on the type and location of the project and its duration, because in some situations, a prudent contractor does not want to be entirely dependent on subcontractors to carry out key operations.

This mix of control complicates and makes the job of the front line supervisor even more difficult. 'The feeling of Company membership and job commitment becomes irrelevant in the circumstances', is the comment of Michael J. Bresnen of the Work and Employment Research Group at Loughborough University, who have conducted extensive research into site recruitment and co-ordination of labour.

Perhaps the most telling of their findings is that the major sources of dissatisfaction commonly expressed by both tradesmen and supervisors are factors such as delays and reworking attributed to inadequate managerial, technical and administrative support in the field.

This confirms the statement made above that no supervisor can perform satisfactorily without the necessary resources. But even with these resources, the task of the supervisor on a construction site is by no means easy.

Time is frequently the scarce commodity, but the skilled supervisor will know that performance cannot be judged solely on a basis of time. Quality is an equally vital factor and more demanding on supervisors than time. If a planned operation time is improved on, this must be of benefit. But with quality, higher standards are often only achieved by doing more work and/or taking more time and are not necessarily of benefit to the project.

The six M's that constitute the resources required for construction are:

- *Money*
- *Minutes*
- *Manpower*
- *Machinery*
- *Materials*
- *Management.*

The priority and mix of these resources varies from job to job. All are important, but time often comes at the top of the list.

Supervising professional staff

The professional employee, for example the design engineer, represents a group that has the characteristics of both supervisor and worker, but also distinct traits of its own. The main distinction is that the standards for his work are set down by the profession.

The Rules for the professional conduct of the Chartered Civil Engineer are set out in the Royal Charter, Bylaws, Regulations, and Rules of the Institution of Civil Engineers.

He cannot be supervised in the way a first line supervisor oversees his team. The foreman can relate more closely to the skill and the manner in which the civil engineer does his work, firstly because, being a manual operation it is clearly visible, and secondly, because he is supervising a more restricted spread of skills.

Case 19.

Three stonemasons were asked what they were doing.

The first replied 'I am making a living.'

The second kept on hammering while he said 'I am doing the best job in the entire country.'

The third said, simply 'I am building a cathedral.'

The third is the reply of a true manager. He has his objective clearly in mind.

The first will no doubt give a fair day's work for a fair day's pay but is never a manager.

The second is the management problem. Good workmanship must be encouraged but it must be related to the needs of the whole project.

This famous management story illustrates the dilemma of supervising tradesmen.

The professional employee cannot be directed but can be guided, taught and helped. He needs rigorous performance standards and high goals, but how he does his work is, to some degree, his responsibility.

According to Drucker the professional employee has four specific needs:

- his job must be as a professional yet he must make a contribution to his organization and know that he makes one and what it is
- he must have opportunities for promotion
- he must have financial incentives for improved performance
- he needs professional recognition both inside and outside the organization he works for.

Further reading

Bresnen, M. J., Bryman, A. E., Ford, J. R., Beardsworth, A. D. & Keil, E. T. Leader orientation of construction site managers. *J. Constn Engrs*, 1986, **112**, No. 3, September, 370–386.

Drucker, Peter. *The practice of management*. Heinemann, London, 1955.

Fayol, H. *General and industrial management*. Pitman, London, 1969.

Humble, J. W. *Improving management performance*. British Institute of Management, London, 1965.

Huneryager, S. G. & Heckmann, I. L. *Human relations in management*. South-Western Publishing, Cincinnati, 1967.

Koontz, H., O'Donnell, C. & Weihrich, H. *Management*. McGraw-Hill, Tokyo, 1980.

5 Personnel development

Introduction
The possible levels of entry into the construction industry are
numerous. There are in addition various career patterns.

To become a professional civil engineer, for example, it used
to be fashionable to buy an indenture with a practising engineer
and to study on a part-time programme or in private over many
years, at the same time as gaining practical experience. Similarly,
to become a line manager with a contracting company it used to
be necessary to have a trade background and to develop the
necessary management skills by 'rubbing shoulders' with one's
peers.

The accepted pattern now is to have a formal education in a
construction related discipline and then gain an understanding of
the 'trade skills' from experience and observation in practice,
and by attending specialist short courses. Although it is unlikely,
it is still possible to enter the industry as a labourer and become
a professional, but the system nevertheless requires that a formal
pattern of education, training and development be followed.

It has been recognized that the system of formal education
leading to construction oriented qualifications does not necessarily produce good managers. On the whole, engineering graduates
tend to be relatively poor communicators with little appreciation
of the commercial aspects of their work. To try and circumvent
the problem, some contracting companies are experimenting
with the use of arts graduates in line management roles, because
they are, largely, better communicators and seem to develop
management skills more easily than engineering graduates. At
the same time, educational establishments are introducing

management as a subject for study by engineering and building students.

Management skills

The importance of the personnel function in construction is at long last becoming recognized. Although mechanical plant and equipment is being increasingly used to replace expensive labour, and while utilization skills and techniques have been developed for such equipment, it is no less important to develop high skills in the management of the fewer, but very costly, people concerned in the construction processes. The definition of the job, the attendant selection of the person to fit the job and the subsequent development and appraisal of that person, are essential if the skills of human resources are to match advances in other directions within the industry.

The purpose of this chapter is to examine the aspects of personnel development which are important in the context of the construction industry.

The subject of personnel development is highly subjective and one not easily put into a valid scientific context that is convincing. The larger and more established organizations within the industry might find the ideas more acceptable and more applicable than smaller groups. It must also be recognized that some sections of the construction industry have no real need to apply the principles of personnel development. Many large contractors, for example, carry out work using subcontracted labour employed on a short-term, contract by contract basis. Although good working relationships can be established with time, the main contractor has no obligation, neither contractual nor moral, to develop the skills of these people. They should, however, recognize the need to apply the ideas to their own staff and key tradesmen.

Manpower planning

Manpower planning at the corporate level is a complex task involving the balancing of a range of skills over long time periods, within the context of uncertain market requirements and financial constraints. Successful manpower planning is linked very strongly to having the right people in the right place at the right time, a task which is much easier said than done.

Manpower planning cannot exist in isolation. It is part of the

---

Case 20.

An analysis carried out by E. F. L. Brech showed that, for all industry in the UK, the ratio between operatives and non-operatives fell in the range of 4/1 to 8/1. A subsequent analysis of UK construction companies carried out by D. A. Barratt (1974) revealed that the most profitable organizations had a ratio of six to one and total company employment of 35.

See Brech, E. F. L. *Construction management in principle and practice.*

---

whole process of management planning, concerned with planning the requirement and supply of human resources, as distinct from financial resources, equipment and materials. These human resources include the 'intellectual property' of the company, which is easily lost or misused and needs careful planning and 'nursing' care.

Within the process of business, a large range of assumptions will be evolved, and manning requirements will be related to these assumptions. Stemming from these calculations on manning will be the design of the organization.

For management positions, manpower ceases to be a matter of numbers by category and becomes linked to individual positions and individual incumbents.

Inventory

Making an inventory of the workforce and assessing the rate and form of change are essential aspects in determining future management needs, since what cannot be provided from within must be sought from other sources. For completeness, this inventory should include performance appraisal, potential and psychological profiles.

In building up the inventory data base, employment data can be coded for ready access and analysis. Data on absenteeism and overtime form an essential part of this information system.

The basic inventory should cover a head count, information on qualifications, training and experience, deployment, an analysis of age distribution and an appraisal of usefulness to the organization. Contingency plans to cater for the unexpected form part of manpower planning.

Employee involvement is important and data in the hands of

supervisors, shop stewards and managers can provide useful standards on which to base estimates. With the co-operation of the workforce, job satisfaction can be raised in quality and absenteeism and manpower loss can be reduced.

Human assets

Whereas machines can be purchased or hired, scrapped or reconditioned, cannibalized or transferred to new locations at the discretion of management, human assets cannot be treated in this way. Management ought to have the responsibility to persuade and encourage acceptance of change in an environment of uncertainty and, in the process, should not forget that most employees consider that they have the right to work. Unfortunately, human resources in the construction industry are not always given sufficient respect.

Much of the reaction of organized labour in the past can be recognized as a defence against the lack of fairness and as a striving for extending the rights of employees. Sensibly therefore, manpower planning should take due account of trends in industrial relations policies.

It should be borne in mind at all times that a company is only as good as the people it employs and its future aspirations depend very strongly on the way that those people are planned for, organized and nurtured. Even though most contractors and some consultants are moving more towards subcontracted and temporary labour, key employees would benefit and respond positively to a programme of development.

Job definition

Organization of people to form teams for the performance of specific tasks is essential; it is a disciplined framework within which people can work efficiently. Individuals, once they are in the team, are all important. The manager's task is first to define the job to be done and then to select the person, whether this be from:

- outside the existing organization through job advertisement, short-listing, interviewing and, finally offering employment

- the existing organization's personnel, where there should be the benefit of regular appraisal reports at intervals of a year, or less, from superiors for whom the person has worked. In this case, selection may be made for short-list interview before the team vacancy is offered to the person finally selected

- a combination of both.

The job specification is fixed with the job in mind. Once a particular person has been appointed to the job, it may be necessary to adapt the job specification to suit the job holder, the reason for this being that we do not work in an ideal world where everything and everybody fits a standard pattern or classification. Each person is unique and must be treated as such, particularly when placing a person in a particular job. We all have our strengths and weaknesses in personality, health and physical characteristics, which may affect our job performance capability.

Strengths and weaknesses

Although it is not always possible, the manager must try to fit the job specification to the particular person, as soon as the strengths and weaknesses of the job holder are known. This may be before the person takes up the job, if the manager knows or has spotted that strength or weakness at interview, or by reference to previous employers or superiors. It may, however, be necessary to adapt the job specification to the person after observation in the early performance of the job. When managers are dealing with people, nothing is fixed and rigid. People are living creatures and are stewards of their talents over a period of time, and as time elapses, so may their strengths and weaknesses change. The manager should be alert to this fact at all times and the good manager monitors the difference between the requirements of the job specification and the work of the job holder at regular intervals and spots the strengths and weaknesses before they adversely affect the overall performance.

Fitting the job to the person is merely refining the particular job specification to suit the particular job holder in order to take advantage of the job holder's strengths and remedy his weaknesses. This is done after thorough observation and monitoring by taking duties off one job holder and adding them to another who is near at hand (perhaps a member of the task team) and better able to perform those particular duties. Similarly, as a person is relieved of certain duties, others may be added. Whatever is done in this respect, the job specifications of the persons involved should be immediately altered to suit the new situation.

Job requirements

Before preparing job descriptions for individuals required to make up a team for a particular operation, it is important to determine the job requirements. This exercise should consider the following.

- The number of persons required to undertake the activity, bearing in mind the volume of work to be done in the time available. The numbers required may vary during the period of the short-term plans which comprise the complete master plan. Of course, the assessment of numbers required will be dependent upon the output forecast of each person during the short-term plan tasks. This output may depend upon the weather conditions, if work is out on-site, or other environmental conditions if indoors.

- The type of personality required, where a person who can mix well and motivate his team into top performance in the various conditions of weather, environment and morale that will be encountered during the task, be it days, weeks or months, or a worker whose task can be performed in isolated conditions.

- The academic and technical qualifications desirable and the length and quality of the job experience that are essential to be able to perform the individual task required.

- The age range, health and physical fitness that are essential for the job holder.

- The detailed duties of the particular job and the output/performance required.

- The precise authority to be given to the person to enable the job to be done. This will involve limiting what is allowed and what is not allowed.

- The designated person to whom the job holder will be directly responsible and what happens in the absence of that person.

- The persons who will come under the direct control of the job holder to enable the tasks to be performed. One person can only have one boss for any particular time and task.

Having designed the requirements of the job, it is essential for these requirements to be set down in an orderly manner under item headings, so that, in the least number of words, the job is clearly defined.

Selection

The report of the Finniston Committee of Enquiry into the engineering profession stated that:

> ... the system required students to commit themselves to a specific type of engineering, when applying for admission (to undergraduate courses) at a time when they can have only a limited appreciation of its various branches or their own aptitudes.

It is acknowledged that people with a national propensity for a particular work situation will perform better than those with no such ability. It follows that people recruited into situations for which they have no aptitude will be the cause of much expense in training and education which might ultimately be wasted. Moreover, the waste is compounded if those people could have been found employment more suited to their natural talents.

Line management in the construction industry requires particular skills and talents which not everyone possesses and, if the aptitude for working in such a situation can be discovered at an early stage, then resources allocated to developing these talents will show a quick return on investment.

The young manager on-site will be concerned with getting the work done on time and within almost impossible financial limitations. The demands of the client, the designer, the contracting company, the workforce, the shareholder; all have to be satisfied. The young manager must therefore learn quickly and methodically how to shoulder his responsibilities.

Tests

According to Lawton in the *Management development and training handbook* (Taylor & Lippitt, 1983) there is no absolutely reliable method for testing and assessing potential of any kind, especially where executive potential is concerned. However, some testing and assessment methods have proved satisfactory, if not completely accurate.

The armed forces have used their own selection methods for many years and many large organizations which can afford such techniques find them valuable in screening applicants for company positions. In fact, some of the larger construction companies have used methods similar to those used by the armed forces, the reason being that the nature of the work, while not military, has certain similarities. Construction line managers need to be good leaders and be capable of working under stress in demanding situations out of doors.

If one were to accept the validity of the work of psychologists, then it is evident that people have different personality types which can be identified using tests of word association. If it were possible to match personality types with the natural aptitudes for construction personnel, then one would have a readily available test for identifying potential.

The terms 'selection' and 'identification' are occasionally confused. Identification of potential is a general process that may involve a number of candidates, whereas selection usually refers to the choice of one of several candidates. Identification ought to occur before selection and should therefore be capable of being applied before the candidate has acquired skills pertinent to the task.

Techniques

Techniques for identifying potential are primarily of two types: traits-oriented or results-oriented. In the former, the person's personal qualities are the prime considerations; in the latter, the person's performance on past and current jobs is the main factor. Most selection programmes combine both types of techniques and it is essential that the correct method be used at the appropriate stage in a person's career. It is useful at this juncture to describe briefly the major tests and assessments for identifying potential.

The techniques practised by assessment centres are used by comparatively large organizations. The basic ingredient of this approach is to expose an employee, in a structured environment, to a variety of exercises, tests, simulated situations, and the like. Reactions are observed and evaluated in terms of potential effective behaviour in a particular job within the particular organization. The centre can be on the site of the jobs for which candidates are being assessed, or remote from the action.

The principles involved in this approach are both traits-oriented and results-oriented and have proved useful ever since their limited use prior to and during the Second World War, and more extensively since in the training of officers and in studies in personality. The use of this approach in assessing manager potential for industry is comparatively recent, and not extensively used due to the costs involved.

The validity of work experience and seniority as indicators of potential are highly questionable. Neither of these is an indicator of effectiveness, yet many organizations depend greatly on a candidate's experience as a determinant in recruiting and promoting employees. While it is acknowledged that exposure to jobs implies at least some acquisition of knowledge and skills, too much weight is given to it when considering employees for new jobs which might entail a different relationship with superiors, peers and subordinates.

The use of inventory forms, which furnish background information on the education, training, experience and other biographical data of the candidate, is one of the easiest and most widespread of all techniques for identifying the potential of employees. They could also be the most misleading. Only too often, the inventory is the sole basis for initial selection in the screening of employees.

Interviews

Interviews are primarily traits-oriented and are, perhaps, the most widely used technique for identifying potential, although they are frequently used with other techniques. The reactions of the interviewer tend to be subjective and patterned in his own image. Even when the interview is conducted by a psychologist or psychiatrist, the atmosphere may not be conducive to completely objective results.

Performance evaluation is obviously results-oriented but it often reflects the traits of the evaluator rather than the candidate. Where there are no specific criteria for ranking, the results are inconsistent, subjective and consequently misleading if taken at face value. Although many people believe that it is the best predictor, the technique should be used only in combination with others and should be derived from more than one source to be a reasonably reliable indicator of performance.

References and recommendations can be obtained from or made by superiors, peers and subordinates, and from persons outside the employing organization. Whether these are analysed in-house or by outside consultants, they still have the basic defect that they are usually inconsistent and biased, often deliberately so. References are usually supplemented with biographical data.

Tests of one kind or another have been used for selecting leaders since time immemorial. Their variety is immense. Those used specifically to identify management potential include psychological tests, but normally they are used in conjunction with other selection methods. Among the standard tests in use are intelligence tests, mental ability tests, interest and personality tests. They test cognitive, verbal and arithmetic ability as well as personality, interest and projective qualities and, although they are sometimes adapted for particular organizations, they are not normally related specifically to the jobs which are being sought by candidates. Research is growing in this field because the tests are fairly inexpensive to administer.

Potential

Training and development programmes are included here among the tests used in identifying potential because achievement in training programmes may frequently be an indicator to performance on the job itself. This is particularly true where the programme is directly related to the jobs to be filled. There is probably scope in refining the monitoring of student performance on education programmes as an indicator of potential for the work environment.

We have now gone full circle and come back to the fundamental question posed by this chapter; how can we identify potential, especially management potential, in the construction industry? The working environment of the construction industry requires total commitment to the job in hand; half-hearted leadership results in less than half-hearted performance from those controlled. Clearly one is looking for strong traits, possibly linked to the psychological make-up of the individual.

In recent years, changes in economic conditions and in legislation have made the tasks of selection much more involved for the selectors. The Sex Discrimination Act 1975 has forced selectors to be more cautious in their selection procedures. Selection decisions must now show clear association with the performance demands of the job.

Selecting managers

Once policy towards management development has been decided, the next decisions concern the use of various techniques that can be applied to the task of selecting managers. According to G. A. Randell in *Management development and training handbook* (*see* Taylor & Lippitt, 1983) three levels of effort in selection procedure can be classified.

The basic level involves the minimum effort likely to result in selection decisions being made which are better than those that would have been made by chance. This entails a certain amount of effort in understanding the job that needs to be done, and the kind of person who would be best suited for it. This is then followed by interviewing a range of applicants, and offering the job to that person who looks best.

The second level goes beyond the basic mainly in detail, and an attempt is made to collect more information about the organization, the job and the applications. It can make use of many techniques, on the assumption that the more information that can be applied to a selection decision, the more probable it is that a good decision will be made.

The highest level is the systematic method of selection. The systematic level may not use as much data as the other methods, but what it does use is known, through research, to be relevant to the selection decision. Its essence is follow-up information, where the predictors of management behaviour which are used have been checked against measures of that behaviour.

There is a move away from systematic testing back towards the basic kinds of procedures. This is because, even though it is not possible to prove the validity of the more subjective ways of assessing applications, it also follows that they cannot be proved wrong.

It would appear that the construction industry needs to be clear about the types of people it requires. It needs to specify jobs and try people for a test of fit, just as one would specify and test materials or machines for suitability. After all, the human resources are probably the most important aspect of a project.

The following is abstracted from *The manager's guidebook* (Elliot, 1983) and serves as a useful guide to both interviewers and candidates alike.

Notes for interviewers

The applicant should be pleasantly received by a person who is expecting him. He should be seen punctually, and arrangements should be made to ensure that the interview is not interrupted.

Serious and full attention should be paid to him throughout the meeting and he should not be allowed to get the impression that the interviewer would prefer to be doing something else. He should be put at his ease and encouraged to talk freely.

The interviewer should keep his own personality out of the interview as much as possible and should concentrate on understanding the applicant's point of view. When asking searching questions, he should do so in such a way that the applicant does not get the impression that he is being cross-examined.

The interviewer should have a thorough knowledge of the job requirements, and should be sincere. He should avoid meaningless conversation and should ask only questions that are designed to produce information that will contribute to his understanding of the candidate. He must remember the objectives of the interview:

- to determine the relevance of the applicant's training and experience to the requirements of the job under consideration
- to appraise his personality, character and motivation
- to compare the applicant's suitability with that of the other candidates.

It is quite likely that about half the interview will be a probe into the various jobs that the candidate has had during the immediately preceding 10 years. Searching questions which may be asked about each job are:

- How did you get the job?
- At first, what was the nature of the job and what was the remuneration?
- What progress did you make, and what was your final remuneration?
- How closely were you supervised?
- How did you get on with your immediate superior and with your colleagues?
- What did you achieve, in measurable terms?

- What did you especially like and dislike about the job?
- Why did you leave?

The candidate will also be considering whether he really wants the job offered. The better the candidate, the more likely it is that he will weigh up any offer against other companies. The interviewer should decide before the interview what he intends to tell the candidate about the company and the job and prepare to put it across effectively.

A formal written system of assessment helps the interviewer to be more objective. The following plan provides a suitable framework. It is important, however, to consider the relevance of each point to the job and to give suitable weightings.

- *Knowledge and achievements.* What type of education has he had? How well has he done educationally? What occupational training and experience has he had already? How well has he done occupationally?

- *General abilities.* Has he any marked mechanical aptitude? Manual dexterity? Facility in the use of words? Or figures? Talent for drawing? Or music?

- *Appearance, impression.* How agreeable are his appearance, his bearing and his speech?

- *Interests.* To what extent are his interests intellectual? Practical or constructional? Physically active? Social? Artistic?

- *Physical capacity.* Has he any defects of health or physique, which you can observe, that may be of occupational importance?

- *Opportunities and circumstances.* What are his domestic circumstances? Is his family a help or a hindrance? Does he own his house? How mobile is he? Is he pursuing a course of study?

Under each of the above six headings, one of the following ratings should be given:

A = in top 10% of the population   Outstanding
B = in the next 20%   Good
C = in the middle 40%   average
D = in the next 20%   Below average
E = in the bottom 10%   Unsatisfactory

Induction and training

The first few weeks spent in a new job or with a new organization are vitally important. It is during this period that some permanent positive or negative attitudes are created in the mind of the newcomer. It is, therefore, of some considerable importance that we set out to ensure that our new recruit is absorbed into the organization effectively.

Essentially there are two parts to this process. First, the initial introduction and broad induction which lasts for two to three days, and second, the period of settling in, briefing and training which can extend over any period from one or two days to many weeks depending on the level of person recruited.

What is absolutely necessary is the planning of these periods. Carried out thoughtfully, such induction and post-appointment planned experience and training pay dividends.

The question 'whose responsibility is it to arrange such matters?' can be answered in different ways.

Induction

Induction can be the responsibility of personnel or training departments, or that of line management. In the event, it will depend upon whether the new recruit is a single appointee or one of a batch, upon the level of post being filled, and upon the function or department.

Line management should perhaps take the lead in this. If we do regard human resources as capital assets, then the line manager will want to ensure that he is getting return on the capital invested and he will make sure that his new 'boys' get off to a flying start.

Introductory sessions and induction courses have already been introduced throughout the industry for younger new starters. Experience shows that clear benefits have accrued from these and where, experimentally, a more mature entrant has been included in the process, the reaction from him has been very encouraging.

It is very sad that, where improvement in job performance is sought, so often a request is made for a 'course'. Training courses have limited value — they are not the panacea for all ills in a department or a section. Far too often courses are asked for when, in many cases, the cause of low performance is the result of inadequate or over-sophisticated systems, poor human relations, lack of communications or just plain bad management.

Investigation of a suggested training need often pinpoints the real reason for low job performance and in this sense a training department is behaving as a consultant on the ultilization of human resources.

Case 21.

Development is always self-development. Nothing could be more absurd than for the enterprise to assume responsibility for the development of a man. The responsibility rests with the individual, his abilities, his efforts. No business enterprise is competent, let alone obligated, to substitute its efforts for the self-development efforts of the individual.

But every manager in a business has the opportunity to encourage individual self-development or to stifle it, to direct it or to misdirect it.

P. Drucker, *The effective executive.*

Training

The word 'training' really means providing the conditions, the instruction and the encouragement which assist the learning process. The basic objective of good training is to help people to learn and since this is a personal desire they cannot be forced to do so.

Good management creates an environment which encourages people to question, criticize and acquire knowledge. Formal training courses can play a part but they are secondary to the development of an attitude of receptiveness.

It is exceptionally important to recognize that one of the major responsibilities of line management is to encourage development and to provide the opportunity for learning. These must align with the objectives and needs of the organization.

Over the past five years, increasing attention has been paid to 'on the job' training. This has meant that managers and supervisors have had their attention drawn, often unwillingly, to their responsibility for improving job performance and providing the environment in which it will occur.

Help in this field has been sought from training personnel and, undoubtedly, more attention will be given to providing line managers with an understanding of teaching and learning processes and of training techniques.

Performance appraisal

It is essential to have an annual appraisal of each member of staff for the purpose of policy decisions for:

- a decision on whether the person is of sufficient ability and capability to be trained for future career development
- a decision on whether the person is to be retained in the short-term due to specialist knowledge or particular job requirement
- a decision on whether to retain the person at all
- a decision on how the dismissal can be carried out without being involved in a court case for wrongful dismissal
- the purposes of the annual salary review

The problems following assessment are to face the fact that action must be taken on the decisions reached. It must be realized that personality problems and square pegs in round holes do not improve with time. It is a fact they generally get worse and become a bigger problem, the more time you leave them without taking action.

In concluding this section, it should be appreciated that the end product of a performance appraisal is a contract between the employer and the employee. To this end, the employee has certain obligations which he must fulfil in order to progress along the prescribed route. By the same token, the employer must provide the necessary resources and facilities which will enable the agreed objectives to be achieved. As with any normal contract the conditions, rewards and penalties must be clearly understood by both parties.

Management development systems

Staff development is concerned primarily with improving a person's performance in work — it is wasteful and unrealistic to develop people without reference to the organization's objectives and the needs which arise therefrom. It follows, therefore, that staff development is carried out to assist people to improve their job performance now and to prepare them for tomorrow. A valuable spin-off is, of course, the build-up of a supply of people for promotion.

An effective appraisal system is the foundation for a good staff development scheme. The gap which exists between present and required job performance should emerge at appraisal, and the steps which should be taken to eliminate that gap can then be identified. They do not consist solely of 'allowing him to go on

THE PAST YEAR   Name:

Position Title   Grade   Company   Dept

Actual projects and assignments in past year. Comments on performance

---

Performance in past year   Comments

Technical ability and know-how

Presentation skills: writing, drawing, etc.

Amount of useful work produced

Technical creativity and ability to motivate

Internal relationships: with colleagues, etc.

Taking the right amount of initiative

External relationships: with clients, etc.

Administration and paperwork

Management ability: planning

Management ability: leading

Management ability: organizing

Management ability: controlling

Sales aptitude

Space for any special comments on performance:

Name of Manager: Date:

Fig. 11.   Sample staff appraisal form

DEVELOPMENT ACTION

In the short-term, what should be done? What knowledge, skills,
technical or managerial know-how, or interpersonal skills would
help to improve performance in the present job?

What are your views about his/her potential development over the
next two to five years? OR What are the main things about his/her
value to us that don't show in the ratings on the previous page?

Comments by your superior

Completed by: Superior:

_____   _____

Fig. 11 — continued

THE JOB-HOLDER'S VIEWS

What are the job-holder's views on his/her own immediate training
and development needs?

In what ways would he/she like to see own career develop in the next
two to five years?

What obstacles, including own abilities, tend to stand in the way
of these plans?

This page completed by:

Fig. 11 — continued

a course'. Many methods are available for development, including on the job coaching, guided reading, undertaking projects, attending internal and external courses, visits and job rotation. The individual's needs will always determine his particular training plan.

Responsibility

The personal responsibility of the manager for identifying and, to some extent, satisfying the training and development needs of his staff is clearly apparent. His role, however, must be supportive and must allow the individual scope for self-development.

Management development schemes have been introduced throughout the industry during the last few years and results are beginning to flow, but only slowly. This is not unexpected because the exercise is extensive and the very nature of the schemes causes there to be an in-built time lag. One could expect the benefits arising from such activity to show much more clearly after two to three years.

The development of staff resources is a neglected part of short- and long-term personnel management. It is one thing to acquire the right calibre of staff at the required time for the performance of an organization's activities. It is another and very important issue to stimulate the development of such staff in order to provide for future requirements, both of the individual members of staff and of the organization. Good staff will not be motivated and retained unless they are given good long-term career prospects.

Continual process

Following the minimum period of annual appraisal, and at other times in special cases, each member of staff should be considered for development into the next stage of his career. Some staff may be satisfied to plod along without much forward development, but all good ones do require continual attention to train them so that they develop into the highest level they are capable of achieving to the advantage both of the individual and the organization.

This development will involve attendance at training courses on specific topics. For engineers seeking a career in management, a proper course of training should be planned, dealing with communication skills, man management, network planning, project time and cost control, financial and commercial business techniques and aspects of law, all of which are essential to the construction manager.

If an organization is not interested in sponsoring the development of the staff they have, the staff will train and develop themselves by moving on to other organizations in order to acquire the knowledge they feel they need in order to progress their career. An organization must train and develop its own staff resources where time will allow. The only other alternative available is to buy in staff from the market place with all its risks of failure and the high costs involved.

Self-management

For a manager to be able to manage a business involving a vast wealth of resources he must first be able to manage himself. Half-hearted leadership and control leads to less than half-hearted performance from those controlled.

One of the aims of any manager of a business enterprise is to ensure a healthy profit, and less than total commitment will result in reduced profits or even a loss situation. In managing himself, a person must pay attention to a number of areas which are important in the creation of an image, the development of self-confidence in self-administration, and in personal development.

The normal working day for most people is about ten hours, including travelling and rest periods. Some managers work longer hours and always seem to be under pressure, dealing with a backlog of work. Others seem to get through an enormous volume of work within a very short space of time.

Ground rules

The most effective managers seem to apply a set of basic ground rules to their working day, which can be summarized as follows.

- Know what jobs need to be tackled. A simple list with short, medium and long-range tasks with deadlines and priorities.
- Don't waste time on non-essential paperwork, discussions and meetings.

- Allocate time to the various tasks in hand.
- Delegate tasks which can be satisfactorily carried out by others.
- Apply the principle of exception* when selecting tasks.
- Allocate time periods for reflection on the overall strategy for dealing with tasks.
- Apply concentrated effort to the execution of jobs and try to avoid interruptions.

*This is an adoption and adaptation of F. W. Taylor's exception principle. This lays down that management reports should go into detail when performance is widely different from past norms or averages — concentrating on the specially good and bad ones. Its use in the context of selecting tasks is to concentrate effort on those tasks which result in positive progress and not to 'waste time on matters that waste time'.

Essential tools in applying these rules are a well kept diary, a clear mind and a good secretary.

Not everyone will be able to apply all of these ideas to their own job. Civil engineers under training, for example, might be

---

Case 22.

According to Hodgson, there are 20 principles of self-development. The more important ones are:

- development is a breakthrough to a new level of potential
- a threshold of difficulty has to be crossed
- an appropriate external challenge is essential
- self-development is self-initiated
- self-development requires self-discipline
- self-development requires learning how to learn about oneself
- primary motivation is through self-achievement and self-fulfilment; external rewards and punishments are secondary
- a self-developer is able to judge whether his intention is strong enough to carry him through
- a self-developer recognizes that nothing will be achieved without personal sacrifice
- self-development requires the guidance of a more mature self-developer

Boydell & Pedler, *Management self-development: current concepts and practices.*

---

working entirely alone in solving engineering problems. They can nevertheless fit the rules to their own situation.

For those on organized training schemes under the sponsorship of an employer approved by a professional body, such as the ICE training scheme for graduates under agreement, the guidelines are dictated by the rules appertaining to the PE examinations. The work experience which supports an application for professional recognition can be conventionally recorded in the personal log-book which should be kept up to date.

Updating

Dependent upon the type of career path to be followed, periods of updating should be included in any self-management programme. For those in engineering design, technical courses to suit the specialist interests should be followed. At least one week per year should be devoted to this. In addition, for those with any inclination for progress into management, the particular skills required in managing people and in finance, legal and contractual matters should be acquired by attending training courses or by private study.

The emphasis should be on identifying the areas of strengths and weaknesses in the personal make-up and reinforcing or eliminating these to suit the choice of career path to be followed. This must naturally be a programme designed to suit the individual and the emphasis has got to be on developing the self by the self.

Development takes place when an intelligent system, in this case the self, breaks through to a new level of potential. This implies some form of change.

The best sources of advice for mapping out a personal career plan are the experiences of others. These can be current or past experiences of peers, employers, acquaintances, and professional advisers and from the documentation of experience to be found in libraries and professional journals.

Further reading

American Society of Civil Engineers. *Personnel: recruiting, motivating, rewarding*. ASCE, Philadelphia, 1983.

Barratt, D. A. *MSc thesis*, University of Surrey, Guildford, 1974.

Boydell, T. H. & Pedler, M. *Management self-development: current concepts and practices.* Gower, London, 1981.

Brech, E. F. L. *Construction management in principle and practice.* Longman, London, 1971.

Drucker, P. *The effective executive.* Heinemann, London, 1967.

Elliot, G. *The managers guidebook*, Kluwer, Brentford, 1983.

Ford, J., Jepson, M., Bryman, A. Keil, T., Bresnen, M. & Beardsworth, M. Management of recruitment in the construction industry. *Project Management* 1, No. 2, May 1983, pp. 76–82.

Institution of Civil Engineers. *Civil engineers for the 1990s.* Thomas Telford, London, 1985.

Mondy, R. W., Holmes, R. E. & Flippo, E. B. *Management: concepts and practices.* Allyn & Bacon, London, 1983.

Taylor, B. & Lippitt, G. *Management development and training handbook.* McGraw-Hill, New York, 1983.

6 Trade unions in construction

Purpose

The main purpose of a trade union is a quite simple one. It is to represent its members in their relations with their employers and to obtain for them the best terms and conditions of employment that it can. If employers were always benevolent and generous, unions would not exist. But, in the real world, not all employers are consistently benevolent and open-handed, and employees often believe that they need the security of an organization to protect them in matters of industrial relations and to bargain with their employers on their behalf.

Few employees believe that the contract between an employer, even a benign one, and an individual employee is a bargain between equals. A union will argue that the employer will always be in a much stronger position than his worker is, and in most cases that must be so. It is only when the employee offers a unique or irreplaceable skill, or occupies a key place in the company or process and has the self-awareness to realize this, that he can meet an employer, or a potential employer, as an equal or even face up to him from stronger ground.

As the employer has the right to hire and fire in the pursuit of his own aims, or those of his owners, and the employee is engaged at a wage or salary in order to supply his labour or his skill — and these are commodities in the world of economics just like any other — their standpoints are different. Thus it is scarcely surprising that their interests differ too, for the most part, and seldom coincide, even if they come close from time to time.

114

But men and groups are often willing to work together, even when they disagree, and they strike bargains which enable them to do so. If the purpose of a union is to represent its members and achieve agreements for them, the method it uses is that of collective bargaining.

There are several hundred unions in Britain, varying in size from the giants, the names of which are household words, to tiny ones which are known only to their members and the employers with whom they deal. About 100 of the most important unions are enrolled in the Trades Union Congress (TUC) which is intended to present the voice and the most attractive face of trades unionism to Government and the public at large. The TUC has no power over its constituent members, but it attempts to reach a concensus view which fairly represents the conflicting opinions which are expressed in its conferences and meetings, and reconciles them.

Collective bargaining

When the TUC was presenting evidence to the Royal Commission on Trade Unions and Employers' Associations (better known as the Donovan Commission) some 20 years ago, it said that collective bargaining was 'in fact more than a method, it is the central feature of trade unionism'. Such a collective bargain might be national in its scope or it might be restricted to a single firm, or one part of a firm, but in every case it is intended to define the terms and conditions of employment of workers as a group and not as individuals. That is why unions put so much stress on unity of purpose among their members and that is why the unions wish all of their members to conform to the decisions of the majority once that majority has come to a decision.

The legal definition of collective bargaining appears in section 32(2) of the Employment Protection (Consolidation) Act 1978, which in turn refers to section 29(1) of the Trade Union and Labour Relations Act 1974. Although there has been a considerable amount of trade union and labour relations legislation since then (and more is expected to come) that definition has survived. In essence, collective bargaining is defined as the process of negotiation in relation to: terms and conditions of employment or working conditions; engagement or non-engagement, or termination or suspension of employment or duties of workers;

115

allocation of work or duties among workers; facilities for union officials; negotiation and consultation procedures including the right of unions to recognition as representing employees.

Before the union can bargain on behalf of employees, however, two important conditions have to be met. The union has, first of all, to have the right to recruit members in a firm and to organize them. Secondly, the firm or industry concerned has to recognize the union as entitled to represent the interests of employees and to negotiate on their behalf.

So the first struggles which a union and its members, or potential members, face are to obtain these rights. Even in the mood of the late 1980s, when unions are less highly regarded than they once were and when their power has been much diminished by recent legislation and the economics of unemployment, most employers are willing to grant these rights willingly enough. Even so, it has to be said that there has been a noticeable increase in the number of non-union firms in the last five to ten years. It is not unreasonable to assume that this trend will continue for some time yet.

Once recognition has been obtained, there is one figure who is prominent in the continuing relations between the union and the employer in the factory, the office or on a construction site. This is the shop steward, an official who has loomed large in the demonology of industrial relations, especially in the eyes of the opponents of unionism.

Shop stewards

When the Donovan Commission reported in 1968, it detected two quite distinct kinds of trade unionism, a formal system and an informal one, co-existing side by side and for the most part in reasonable harmony. The formal system is easily recognizable by its structure of local, branch, divisional, regional and national organs and by its hierarchy of elected or appointed officials corresponding to that structure.

At the same time, as the Commission realized, there is an informal shadow organization which is often very powerful and has, on past occasions, been more powerful than the official union organization. This shadow is manned by the shop stewards, who are essentially the local representatives of the workforce with whom they are in daily contact. The union officers, however, only encounter the members infrequently and generally in conditions of stress or difficulty. It is the shop stewards, and not the union officers, who deal with the day-to-day business of negotiating with management, representing the grievances of their members and attempting to resolve problems as they arise.

Not only is the shop steward the main representative of the members in their relations with management, he is also their main link with the union. For it is largely through him that the policies and attitudes of the union are conveyed to the members, and these are often interpreted or re-interpreted by the shop steward on the way. He also filters the opinions of his members back to the offical union in an attempt to make them aware of the ebb and flow of belief at the shop-floor or on-site.

The popular impression of the shop steward is not a flattering one. He is often perceived as being obstructive and uncooperative. While there may be some truth in this widespread view, it is not the whole truth. When a shop steward is convinced that the management is right on a particular matter, or if he believes that a sticking point has been reached in negotiations, he will usually convey his conviction to his members and try to persuade them to agree with him. But if he fails, it is the members' side which he will take from then on, and not the management's. If it is not his job to be obstructive, it is his business to be awkward.

Negotiations

The trade unions' main way of attaining their objectives is by collective bargaining. Reaching agreement in this way is like settling the final account of a big construction project. The unwritten conventions are much the same in each case.

Each side will argue about such things as economic or financial prospects, the likely costs of any change which might be agreed, loyalty to the firm or to its employees, comparisons with other employees in similar concerns, public policies and constraints, natural justice, precedents in the firm or elsewhere, and almost anything the negotiators can think of. But it is important to realize that neither side is seeking absolute truth.

The union is a pressure group, anxious to gain as much for its members as it can, while the firm is anxious to yield to that pressure to no more than a limited extent. The negotiations are

117

real; they are meant to lead to concessions on each side, and almost always do.

Negotiation is the process by which a conflict is resolved through compromise. If the negotiations are initiated by management, it must have a clear idea of what it is trying to get. If, however, the negotiations are initiated by the union, in a wage or salary claim for instance, or in defence of jobs which are thought to be in danger, management should make sure that it understands the union's claim thoroughly and the reasons which lie behind it, before making any response. Both sides, if they are sensible, will also try to assess their relative strength and the importance of the matter in question.

Management's objectives will usually be fairly simple: to reach an agreed settlement within its financial constraints, and to find ways of recovering any extra costs through improved productivity or other increases in efficiency. It will also want to reach agreement without upsetting any existing rules or conventions of behaviour or performance; without creating any undesirable precedents and without arousing any other groups of employees who might be outside the scope of these negotiations.

The union's aim will be different from the management's, but will be nearer on some issues than might at first appear. While the negotiating process will have strong elements of bluff about it, there will be an imbalance between the two sides which will affect the outcome. The imbalance will depend on the relative weight which the contestants put on the various elements of the discussion.

The union's claim will be designed to solve as many of its current problems as possible, while the managerial response will be intended to solve as many of the company's problems as it can. Both will be aware that it is extremely unlikely that a solution will be found which solves all the problems on each side.

In all probability, the union will have a sticking point some way short of its claim, and management will have a ceiling above which it cannot go. Generally speaking, there will be an area of possible agreement between these two points in which a solution can be found. The business of negotiation is to get the negotiators into that area as quickly as possible, and to settle within it as amicably as can be done.

Settlements can be reached in negotiations when both sides

believe that they have more to gain than to lose by agreeing. However, if they feel they have more to lose than to gain, then the negotiations can deteriorate into disputes and sanctions become possible.

Sanctions

Sanctions can be used by either side. The employers can dismiss people or lock them out; the union members can disrupt normal working by go-slows, working to rule, overtime bans or other forms of non-co-operation. At worst, the union can go on strike and induce other workers to go on strike with them. But all sanctions represent the replacing of argument by a show of strength.

In some countries, the rights of trade unions to bring out their members on strike or to engage in a variety of other sanctions is fairly clearly defined by law. In Britain, another course has been adopted. Instead of defining a right to strike, British law indicates areas within which a union is immune from actions in tort and cannot be sued for damages.

A tort is a civil wrong which can be remedied by damages awarded through the courts or agreed outside the courtroom. A union cannot be sued in its own name for most torts, but it can be sued for civil wrongs involving personal injury or involving the use of its own property. It can be sued, for instance, by people who are libelled in union publications or who are injured on its premises. But if the civil wrong is done 'in contemplation or furtherance of a trade dispute', the union cannot be sued at all. That is the immunity that arises in all discussion of trade union rights, and which is the foundation on which trade union power rests. The extent to which that immunity is justified is central to the current debate on trade union rights and responsibilities.

Definition of immunity

As the report of the Donovan Commission put it, 'the right to strike is, basically, a right to withdraw labour in combination [with others] without being subject to the legal consequences of acting in combination which would, in the past, have followed'.

The immunity for individual union members is defined in section 13 of the Trade Union and Labour Relations Act 1974, which reads in part:

(1) An act done by a person in contemplation or furtherance of a trade dispute shall not be actionable in tort on the ground only:

    (a) that it induces another person to break a contract or interferes or induces any other person to interfere with its performance; or

    (b) that it consists in his threatening that a contract (whether it is one to which he is a party or not) will be broken or its performance interfered with, or that he will induce another person to break a contract or interfere with its performance.

(2) For the avoidance of doubt it is hereby declared that an act done by a person in contemplation or furtherance of a trade dispute is not actionable in tort on the ground only that it is an interference with the trade, business or employment of another person, or with the right of another person to dispose of his capital or his labour as he wills ...

The immunities applying to trade unions are alike to those which apply to individual union members, but are more wide ranging. They are defined in section 14 of the 1974 Act, which reads as follows:

(1) Subject to subsection (2) below, no action in tort shall lie in respect of any act:

    (a) alleged to have been done by or on behalf of a trade union which is not a special register body or by or on behalf of an unincorporated employers' association; or

    (b) alleged to have been done, in connection with the regulation of relations between employers or employers' associations and workers or trade unions, by or on behalf of a trade union which is a special register body or by or on behalf of an employers' association which is a body corporate; or

    (c) alleged to be threatened or to be intended to be done as mentioned in paragraph (a) or (b) above:

against the union or association in its own name, or against the trustees of the union or association, or against any members or officials of the union or association on behalf of themselves and all other members of the union or association.

(2) Subsection (1) above shall not affect the liability of a trade union or employers' association to be sued in respect of the following, if not

arising from an act done in contemplation or furtherance of a trade dispute, that is to say:

(*a*) any negligence, nuisance or breach of duty (whether imposed on them by any rule or law or by or under any enactment) resulting in personal injury to any person; or

(*b*) without prejudice to paragraph (*a*) above, breach of any duty so imposed in connection with the ownership, occupation, possession, control or use of property (whether real or personal or, in Scotland, heritable or moveable).

(3) In this section 'personal injury' includes any disease and any impairment of a person's physical or mental condition.

The Employment Act 1980 appears to have weakened these immunities from legal action in two respects: first, in cases of what is called secondary action — that is industrial action aimed at persons or firms not directly involved in a particular dispute; and second, in cases where an attempt is made to compel workers in a separate firm or at a different workplace from those in dispute to join in the dispute in sympathy. In such cases, injunctions can be taken out against the unions concerned, and if these injunctions are not obeyed the property of the unions can be sequestered. This is a substantial deterrent to secondary action.

Definition of dispute

As trade unions are entitled to immunity from actions in tort only when they are involved in a trade dispute, the definition of a trade dispute is of no little importance. According to section 29(1) of the 1974 Act, the definition is:

In this Act 'trade dispute' means a dispute between employers and workers, or between workers and workers, which is connected with one or more of the following, that is to say:

(*a*) terms and conditions of employment, or the physical conditions in which any workers are required to work;

(*b*) engagement or non-engagement, or termination or suspension of employment or the duties of employment, of one or more workers;

(*c*) allocation of work or the duties of employment as between workers or groups of workers;

(*d*) matters of discipline;

121

(e) the membership or non-membership of a trade union on the part of a worker;

(f) facilities for officials of trade unions; and

(g) machinery for negotiation or consultation, and other procedures, relating to any of the foregoing matters, including the recognition by employers or employers' associations of the right of a trade union to represent workers in any such negotiations or consultation or in the carrying out of such procedures.

This definition has been criticized as going too far in several directions. Notably it has been held to be unfair that unions should not be liable for any damages done to employers caused by inter-union squabbles, and it has also been objected that the phrase 'which is connected with' is too vague. Critics have argued that a trade dispute should be 'wholly or mainly' related to the various items in the clause instead of being merely 'connected' with them.

Working Rule Agreement
The Conciliation Board

In the construction industry, industrial relations are mainly in the hands of the Civil Engineering Construction Conciliation Board (CECCB), certainly as far as manual workers are concerned. The board is very much a going concern and it has made industrial relations in the industry more rational than they once were.

The Board, which was set up originally in 1919 and then reconstituted in 1952, consists of representatives of both employers and employees. As it now stands, employers are represented by the Federation of Civil Engineering Contractors (FCEC) and the employees by three trade unions: the Transport and General Workers' Union (TGWU); the Union of Construction, Allied Trades and Technicians (UCATT); the General, Municipal and Boilermakers Union (GMB).

According to its constitution, the object of the board is 'to establish and maintain amicable industrial relations between employers and operatives in the civil engineering industry with a view to maintaining industrial harmony and avoiding strikes and all other forms of industrial strife'.

The Board distinguishes between the civil engineering and

building industries, saying that these 'are essentially different and distinct', and it undertakes to negotiate wages, hours and working conditions for civil engineering workers and building workers as quite independent groups of employees.

As the Board aims at industrial harmony, it hopes for 'the resolution if possible without strike or lock-out of any dispute or question which may give rise to a dispute in the civil engineering industry'. It would be hard to maintain that the Board, however good its intentions, has quite managed to achieve that lofty aim. Nevertheless, its day-by-day efforts have brought a considerable amount of greatly needed cohesion to industrial relations in a much fragmented industry.

Originally the fragmentation of the industry was reflected in the number of unions participating in the Board's work. In 1919, for instance, no fewer than eight unions were involved. Two of these have since gone out of existence, three have been amalgamated into the GMB and two others have been absorbed by the TGWU. By 1941, the number of unions had dwindled to three, but in the early 1950s, the board was widened to include craftsmen as well as other manual workers, and the number of unions rose to six. Further amalgamations and mergers, mainly associated with the formation of UCATT, led eventually to the present situation in which operatives are represented by three unions only.

The Board functions through two panels, representing operatives and employers respectively. These are made up as follows: an operatives' panel of not more than 13, composed of five representatives of the TGWU, four of the GMB and three of UCATT; and an employers' panel composed of six representatives nominated by the FCEC.

Each of these bodies forms a separate panel of the Board and appoints its own chairman and secretary. At the time of writing, the Secretary of the operatives' panel is George Henderson, National Secretary of the Building Crafts Section of the TGWU, and the Secretary of the employers' panel is Ron Emery, Director of Industrial Relations at the FCEC and a member of the Institution of Civil Engineers.

Votes are virtually unknown at Board meetings, but should one be needed each panel votes separately under its own chairman, and the individual members are bound by the majority decision of their respective panel once it has been reached. In the end, however, decisions are made by the full Board.

Apart from resolving particular disputes as they arise, the Board's main function is to revise the Working Rule Agreement, which defines the nationally agreed rates of pay and working conditions for operatives and building craftsmen involved in civil engineering, and to bring it up to date when necessary.

Disputes procedure

Long before an industrial dispute reaches the Board, however, it will have followed through a procedure which is carefully laid down in the Board's constitution and has since been embodied in a working rule. Among other things, this rule encourages union membership and defines the role of the shop stewards.

If the legalistic phraseology of the rule is stripped away, it means that the Secretaries give such advice as they can before putting a problem to the Board's monthly meeting or to a disputes sub-committee, if they have to do so. Disputes which affect a single site only, ought to be settled at that site, but those which spill over into the industry in general should be resolved by the Board. In other words, the Board is not to be regarded as a referee in every dispute, although both particular and general disputes can come before it if necessary.

Right at the end of the Board's constitution is a provision which is a very important one in British trade unionism. The unions have always been extremely wary of the incursion of law into industrial relations. In particular, they have been unwilling to make legally binding agreements with employers. They have been, and remain, suspicious of the law and lawyers, sometimes for good historical reasons. So the ninth article of the constitution notes that 'this agreement, although not legally enforceable, has been entered into freely and voluntarily and is intended to be binding in honour as between the parties, and the parties therefore undertake to take all such steps as are reasonably practicable to prevent any persons covered by it from acting in breach of its provisions'.

The agreement is meant to be upheld by custom, but the sad truth is that the provisions of the agreement are most likely to be upheld by both sides when it suits them to do so, and only then. The Working Rule Agreement is somewhat complex and runs to

23 rules and some 50 pages. It lays down the rates of pay for employees in detail, making special distinctions for those working in the vicinity of London and certain parts of Merseyside. The agreement stipulates plus rates on the general basic rates for operatives for various specified skills, and indicates the construction industry training schemes applicable to acquiring those skills. Though the operative usually has to show proof of training and experience, the onus of checking that he has the skills he claims is on the employer.

The agreement also stipulates the minimum weekly guaranteed bonus to be paid and the travel and subsistence allowances to which workmen are entitled. There is a variety of allowances which depend on working conditions, such as working in tidal water or at a height above ground or exposed to severe weather conditions. Special allowances can be awarded by the CECCB for work done in conditions which are 'dirty or obnoxious or rigorous to a degree in excess of the conditions inherent in, and normally encountered in, the industry'.

The working rule agreement is thus comprehensive, and it covers virtually every possible area of conflict between civil engineering employers and their workmen. It can be ignored only at the peril of disruption and the probable stoppage of work.

Safety

The construction industry is one of the most dangerous in Britain. According to George Henderson, Secretary of the Building Crafts Section of the Transport and General Workers Union (the biggest union in construction) there are, on average, two fatal accidents in the industry every four working days. Consequently, site safety is uppermost in the minds of the unions in the construction industry, although it has to be said that this is not always true of the operatives themselves, especially those who are engaged in labour-only subcontracts.

The Working Rule Agreement, therefore, deals in considerable detail with the requirements of the Health and Safety at Work Act 1974 and provides specific notes for guidance on numerous aspects of safety procedures and behaviour. The TGWU published a comprehensive guide to the Act for the use of its members who were, or were likely to become, safety representatives on sites under the requirements of the Act. The union's intention is to give general advice on how to make sure that the provisions of the Act become effective in preventing accidents on-site. The union also runs educational courses for its members and maintains a health and safety officer in each of its eleven regions.

Construction unions

There are several trade unions which operate and recruit in the construction industry. Obviously, most of their members are in the building and construction trades rather than in the professions, though a fair number of construction professionals are members of the National and Local Government Officers Association, the much smaller Federated Union of Managerial and Professional Officers, and similar bodies. While the Union of Construction, Allied Trades and Technicians has a membership of over 200 000, including a substantial number of architects, engineers and planners who are organized in the Supervisory, Technical, Administrative, Managerial and Professional Section of that union, the biggest and most powerful union in the industry is the Transport and General Workers' Union, the Building Crafts Section of which claims a membership in the construction and allied industries of over 250 000.

The TGWU claims a large membership, including bricklayers, carpenters, concrete finishers, crane drivers, dumper drivers, fitters, lorry drivers, painters, pipelayers, plasterers, ready-mix concrete drivers, scaffolders, steel erectors, steel fixers, stone masons, tilers, tunnel miners, and other ancillary workers. It is clear that the union is widespread throughout the industry. In addition, however, TGWU members provide building materials, produce machinery used in the industry and transport materials and machinery to sites.

The TGWU insists that the union belongs to its members, and always has done so. Each member is attached to one of the union's 8000 branches, and it is in these branches that union policies are said to originate. That may well be true in part, but in a giant union like the TGWU it is hardly likely to be wholly true. The main source of power in the union is likely to be at the top or very near it, no matter how involved members are at their local level. Still, the union's structure must be described as a democratic one, and its members have a bigger say in its affairs than the average shareholder takes in the company he partly owns, although, of course, a company director has legal obligations to the shareholders.

The union has an elaborate structure of 14 trade groups and 11 administrative regions with a system of national, regional and district committees which supervise the union's activities and co-ordinates such things as wage claims and other demands on employers. The union's supreme authority is its biennial delegate conference, and between its meetings the union is run by a general executive council which is elected partly by the regions and partly by the trade groups.

It is important to realize that democracy in the TGWU, as in all parts of the wider labour movement, is representative democracy. But it is representative democracy with a difference. While, for most purposes, the representative will be left alone once he is elected, apart from reporting back to the minority of his constituents who are likely to turn up at branch or other meetings, that is not always so.

In cases of dispute with an employer, a proposed settlement has to be approved by the members who are involved in the dispute. At one time that was done by means of mass meetings, but under current legislation has to be done by means of secret ballots, and will probably have to be done by secret postal ballots before long. While the mass meeting was never a wholly satisfactory way of dealing with such matters, it was, nonetheless, a historic method of participation in public affairs, and its use indicates that, while democracy in the unions has never been perfect, a democratic way of working has always been behind the attitudes of trades unions, however imperfect and distorted it may sometimes have been.

Further reading

There is a vast amount of literature on industrial relations and trade unions and their practice and history. In recent years, the law relating to industrial relations has been changing rapidly. Consequently, books become out of date very soon after being published. A small number of relevant books, mainly published fairly recently, is listed below.

Bamber, G. *Militant managers*. Gower, Aldershot, 1986.

Brown, W. *The changing contours of British industrial relations.* Blackwell, Oxford, 1981.

Gallagher, T. *Industrial relations on site.* Construction Press, London, 1984.

Hawkins, K. *A handbook of industrial relations practice.* Kogan Page, London, 1979.

Howie, W. *Trade unions and the professional engineer.* Thomas Telford, London, 1977.

Howie, W. *Trade unions in construction.* Thomas Telford, London, 1981.

Khan, P. *et al. Picketing.* Routlege & Kegan Paul, London, 1983.

MacFarlane, L. *The right to strike.* Penguin, London, 1981.

McMullen, J. *Rights at work.* Pluto Press, London, 1983.

Palmer, G. *British industrial relations.* Allen & Unwin, London, 1983.

Royal commission on trade unions and employers' associations. (The Donovan report.) Cmnd 3623. HMSO, London, 1968.

Biographies

A. S. Martin, MSc, CEng, FICE, FIHT, FBIM

Until taking early retirement in 1985, Stanley Martin had been a chief officer for some 30 years, serving three local authorities, culminating in the post of Director of Technical Services of Erewash Borough Council, where he was responsible for engineering, architecture, town planning and recreation.

He has been active in the affairs of several institutions and associations, both locally and nationally, over many years, and among other things was President of the Association of Chief Technical Officers, Chairman of the Association of Public Service Professional Engineers, and of the Federation of Professional Officers Associations, as well as Chairman of the Membership Committee of the Institution of Municipal Engineers.

A keen student of management, he was awarded the degree of MSc by Loughborough University of Technology for research into aspects of municipal engineering management. He has written over 100 data sheets on management, the first 79 being published in *Can You Manage?* (Municipal Publications, 1981).

He is an examiner for the Institution of Civil Engineers' examination in Management and Public Administration, and is a regular contributor to *Municipal Journal*. He is a member of the Association of Municipal Engineers' Affairs and Editorial Committees, and is the Engineering Council's nominee on EGC 3. He is a Freeman of the City of London.

F. Grover, BSc(Eng), CEng, ACGI, FICE, FIHT

Frederick Grover is a Chartered Engineer with substantial experience in management relating to the civil and mechanical engineering industries in a number of countries.

After obtaining a degree in civil engineering at Imperial College, and war service in the Royal Navy, he spent some time on construction projects in South America and the UK. He has had 15 years experience at director and managing director level, in the management of companies in construction and allied mechanical engineering.

H. K. H. Claxton, MA, MBA, MICE

After graduating from Oxford, Humphrey Claxton spent six years with Freeman Fox and Partners, the last three of which were in Istanbul, working on the Bosphorus Bridge.

He then joined Redpath Dorman Long and worked on bridge construction sites as subagent, and then agent, until he went out to Lagos as Technical Director of Dorman Long in Nigeria. After two years there, he took an MBA at Cranfield, from which he joined Paterson Candy International, becoming Operations Director in 1984.

Currently, as Corporate Development Manager for Portals Water Treatment, he is working on the formulation and improvement of strategies aimed at enhancing business performance.

His greatest enthusiasms are for strategic planning, achieving results through people, and managing improvements in organizational performance.

D. A. Barratt, MSc, DipCE, MICE, MIWES, FGS

David Barratt entered the construction industry at the age of 17 in 1961, when motorway construction was flourishing. During his career he has worked on motorway projects, heavy foundations, and in water engineering, both in the UK and overseas.

He joined the construction department of Oxford Polytechnic in 1978 where he specializes in construction management. He is currently involved in research and consultancy in human resources development.

K. J. Hayzelden, BA, FIPM, MITO

K. J. Hayzelden was educated at Peter Symonds School, Winchester, and Birmingham University where he received a BA degree in Spanish, and at Aston University where he qualified as a member of the Institute of Personnel Management. He is an Affiliate of the Chartered Institute of Building.

He began his career with the Essex County Council Education Committee and then spent four years as training officer with GEC, before joining Robert M. Douglas Holdings as Manager – Group Training Services in 1966.

He is a member of the Construction Industry Training Board's Civil Engineering Committee and its CCSMST Committee, as well as being on the Federation of Civil Engineering Contractors' Training Committee. He was formerly a member of the Lighthill Committee.

J. V. Tagg, CEng, FICE, FIStructE, FRICS, MRTPI, FCIArb, FIHT, MConsE

James Tagg is Managing Director of W. S. Atkins & Partners (Midlands) as well as being a director of the main Board. He began his career with Harold Taylor, consulting engineer, and after serving two large local authorities, designing and constructing roads and sewers for six years, he joined Peter Lind & Co., working on power stations and other large projects.

After a spell with Holloway Bros (London), he became Deputy Chief Estimator with W. & C. French & Co., dealing with motorway, sewage works and waterworks tenders. For seven years he was Chief Civil Engineer with Turriff Construction and was latterly also a director of the company. He then ran his own consulting engineer's practice for two years before joining his present firm in 1973.

He was Vice-Chairman of the ICE's Midlands Association in 1976, territorial member for that Association for three years, and has been a member of the Institution's Engineering Management Group Board since 1983. He is a Bayliss Prizeman of the Institution.

Lord Howie of Troon

Currently Director (Internal Relations) of Thomas Telford Ltd., the Institution of Civil Engineers' publishing company, he originally trained and practised as a civil engineer. From 1963 to 1970 he was Member of Parliament for Luton and held a number of posts in the Government's Whips' Office before becoming Vice-Chairman of the Parliamentary Labour Party. Shortly after leaving Parliament, he became public affairs correspondent on the magazine New Civil Engineer, and he continues to combine that post with his present position. In addition, Howie is the

131

trades union advisor to the Institution, and he has published two short books on trade unionism, with special reference to professional engineers and to the construction industry.

A former President of the Association of Supervisory and Executive Engineers, he is President of both the Association for Educational and Training Technology, and the Independent Publishers Guild, and is an advisor to the Council of Managerial and Professional Staffs. Lord Howie is now Pro-Chancellor of the City University.

Lord Howie was educated at Marr College, Troon and the Royal College, Glasgow (now Strathclyde University). He holds a BSc degree and DipRTC in civil engineering and is a Fellow of the Institution of Civil Engineers and of the Royal Society of Arts. He is also a Member of the Société des Ingenieurs et Scientifiques de France. Lord Howie was a member of the Finniston Committee of Inquiry into the engineering profession.

He is also a member of the Worshipful Company of Engineers and a Freeman of the City of London.

In addition to numerous articles and papers, Lord Howie's publications include the following: *Public sector publishing*, 1968 (joint author); *Trade unions and the professional engineer*, 1977; *Trade unions in construction*, 1981; and *Thames Tunnel to Channel Tunnel*, 1987 (joint editor).